Signals and Boundaries

Signals and Boundaries

Building Blocks for Complex Adaptive Systems

John H. Holland

The MIT Press
Cambridge, Massachusetts
London, England

First MIT Press paperback edition, 2014

MIT Press books may be purchased at special quantity discounts for business or sales promotional use. For information, please email special_sales@mitpress.mit.edu.

Set in Sabon by Toppan Best-set Premedia Limited. Printed and bound in the United States of America.

Library of Congress Cataloging-in-Publication Data

Holland, John H. (John Henry), 1929–
Signals and boundaries : building blocks for complex adaptive systems / John H. Holland.
p. cm.
Includes bibliographical references and index.
ISBN 978-0-262-01783-1 (hardcover : alk. paper)—978-0-262-52593-0 (pb.)
1. Adaptive control systems. 2. Adaptation (Biology)—Mathematical models. 3. Signals and signaling—Mathematical models. I. Title.
TJ217.H644 2012
003—dc23

2011052776

10 9 8 7 6 5

Contents

Preface

Many of the ideas in this book grew out of interactions at the Santa Fe Institute and at the University of Michigan's Center for the Study of Complex Systems. Equally important was Bill (S-Y) Wang's re-awakening of my interest in language acquisition and evolution.

In the early 1960s, Bill and I worked together at the University of Michigan, designing the first courses for the Communication Sciences Program (later the Department of Computer and Communication Sciences). After Bill left Michigan for a distinguished career at U.C. Berkeley, we were out of contact for almost 40 years, though I was well aware of Bill's research. One day, a few years ago, I received an email from Bill (then at the City University of Hong Kong after retiring from Berkeley) inviting me to come to City U. to explore ways in which our ideas, combined, might provide a new approach to language. There followed, in short order, a whole series of interactions based on extended meetings in Hong Kong and working groups at the Santa Fe Institute.

It was through Bill that I met Professor Helena Hong Gao and became aware of her remarkable work on children's acquisition of their first action words. Helena is fluent in Chinese, Swedish, and English, so she was able to compare acquisition of action words in those three languages. She discovered the striking fact that in all three languages children acquired their first action words only after they could do the actions those words named. The link between gesture and speech was unmistakable. Through continued interactions, Helena and I began to explore what pre-primate abilities might lead to the increasing levels of consciousness involved in language acquisition.

A later workshop at the Santa Fe Institute, followed by workshops and meetings at the University of Michigan, led to a manifesto by a group of well-known linguists proclaiming that "language is a complex adaptive system." Because I had been exploring the notion of niche in ecosystems, along with the effects of boundaries in biological cells, these interactions drove me to consider the combined effects of idiolects, dialects, and the like, in providing "boundaries" ("them" and "us") between speakers. It was natural to jump from these considerations to looking much more closely at the effects of signals and boundaries in all complex adaptive systems. This book is the result.

If you had asked me when I was 60 if I would still be actively teaching at 80, I would have said "Absolutely not." But here I am, at 83, still teaching at the University of Michigan, and enjoying it. It is not through necessity, but because of the huge benefits of contacts with new students full of curiosity and original insights. And I am extremely fortunate to still have active interactions with colleagues all over the world. It is indeed a complex, intriguing planet!

John Holland
January 2012

1 The Roles of Signals and Boundaries

Ecosystems, governments, biological cells, markets, and complex adaptive systems in general are characterized by intricate hierarchal arrangements of boundaries and signals.

Ecosystems, for example, have highly diverse niches, with smells and visual patterns as signals. Governments have departmental hierarchies, with memoranda as signals. Biological cells have a wealth of membranes, with proteins as signals. Markets have traders and specialists who use buy and sell orders as signals. And so it is with other complex adaptive systems. Despite a wealth of data and descriptions concerning different complex adaptive systems, we still know little about how to steer these systems. It is the stance of this book that widely applicable answers to questions about steering complex adaptive systems can be attained only by studying the origin and the coevolution of signal/boundary hierarchies, much as the intricacies of ecosystems can be understood only by studying the origin and the coevolution of species.

Though the book makes substantial use of data and examples, its objective is not to produce models that make specific predictions from data. Instead, its objective is to examine mechanisms that underpin the development of hierarchies of signals and boundaries in complex adaptive systems. The critical role of mechanisms is nicely illustrated by Jacob Bjerknes' discovery of fronts as a weather-generating mechanism. Before Bjerknes, weather was predicted by statistical trend analysis based on snapshots of current conditions. (For example, the prediction "tomorrow's weather will be like today's" is right about 60 percent of the time.) Bjerknes' discovery suggested where to look for new relevant data, and how to make predictions conditional on those data. Similarly, the objective here is to find mechanisms that suggest where to look for new data that explain the development of signal/boundary hierarchies. The book's ultimate goal is to tie these mechanisms into a single overarching framework that suggests ways to steer complex adaptive systems by modifying signal/boundary hierarchies.

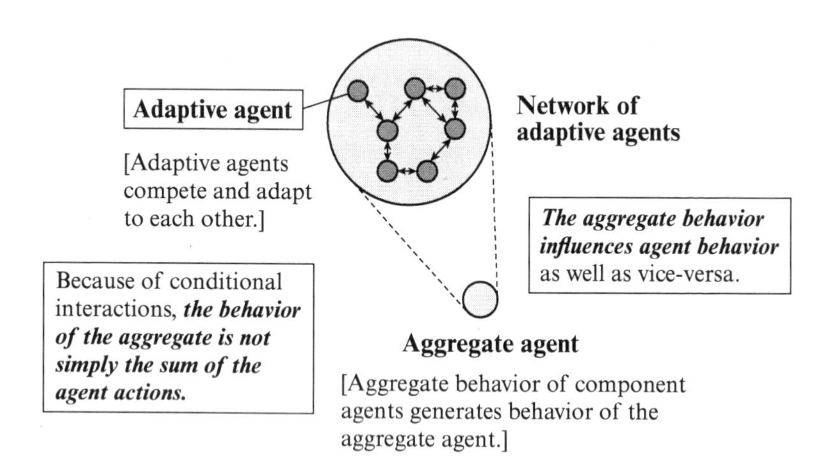
Adaptive agent
[Adaptive agents compete and adapt to each other.]
Network of adaptive agents
The aggregate behavior influences agent behavior as well as vice-versa.
Because of conditional interactions, the behavior of the aggregate is not simply the sum of the agent actions.
Aggregate agent
[Aggregate behavior of component agents generates behavior of the aggregate agent.]

Figure 1.1
Complex adaptive systems.

1.1 Prologue

Consider three events:

• A monkey gives the "eagle-warning" cry, though no eagle is present, watching her companions depart the forest canopy while she remains to feast on the tasty green shoots.

• India and Pakistan contest the dominion of Kashmir, a vale of mountainous terrain that separates them.

• Witches, singing "Double, double, toil and trouble; Fire burn and cauldron bubble," watch over a vat full of exotic organic compounds.

Though these three events seem to have little in common, there is a thread that connects them: in each case, signals, boundaries, and the interactions between them play a critical role.

Consider the monkey's cry. Clearly it is a signal, but what are the boundaries? We must look more closely at the monkey's environment to see the relevant boundaries. This monkey (a vervet) lives in a savannah forest, eating young shoots that grow high in the trees. The savannah forest consists of a variety of resources and nutrients, ranging from leaves on the trees and insects in the bark to the monkeys themselves. We say that the vervets live in a *niche* that consists of the things they eat and the things that eat them. The niche, then, is made up of physical and virtual boundaries that determine the limits of these interactions. By using well-established signals in a deceptive way, the monkey gains a larger share of the resources in its niche. The invisible boundaries that define niches are a complex topic, still only partly understood. And there is another mystery here. Warning cries are pervasive in the animal world, but the animal issuing the warning cry immediately calls the predator's attention to its presence. It is at least plausible that gene combinations that encourage the warning response would die out because of increased predation. What are the mechanisms that encourage the elaboration and spread of warning responses despite this "load"?

Now consider the second example, that of countries contesting territory. Boundary disputes occur throughout the world and persist for decades. To understand political boundaries—even obvious physical barriers, such as China's Great Wall or Hadrian's Wall in Britain—we must examine their origins and their maintenance. Constructing a large wall or a castle, or enforcing a political boundary, requires an extensive social network. To understand the existence of a wall or a national boundary, we must uncover the social and physical mechanisms that make the wall or the boundary possible. The Chinese social network that produced the Great Wall had important similarities to, and some important differences from, the Roman network that produced Hadrian's Wall. Both of these networks also had similarities to, and differences from, the networks that gave rise to feudal castles. In all three cases, there were unanticipated local effects, such as trade and fairs at the portals. There is a kind of semi-permeability, reminiscent of membranes controlling the flow of resources into and out of biological cells. Can we find an overarching evolutionary framework that encompasses the origin and the effects of such different boundaries?

Walls and other political boundaries show up on maps, but what signals are involved? Here we enter the arcane world of diplomacy. Some of the simplest signals, such as moving troops to a border, recall the monkey's bluff. The Cold War was modeled using a simple two-person game called the Prisoner's Dilemma

(Axelrod and Cohen 1999), about which more will be said later. In that game, there is a strategy called "tit-for-tat" wherein one player's current play copies the opponent's previous play. In effect, a player's action signals future intent to do unto the opponent whatever the opponent has previously done unto the player. Many actions in the Cold War similarly signaled future intent, up to and including the "mutually assured destruction" strategy. Of course, the subtleties of diplomatic interaction are usually much more complex. Diplomacy uses language in a complex signal-exchange network, setting boundaries ranging from trading restrictions to territorial rights. Language is arguably the most intricate signaling device on earth, with a biological cell's protein-based signaling networks close behind. Predicting and modifying these interactions usually requires knowledge of the long-term coevolution of the relevant signals and boundaries.

What about the witches' brew? It serves as a metaphor for a complex mixture of chemical reactants and reactions. In the vat, boundaries form where there weren't any before: little "greasy globs" float to the surface, or there's a precipitation of "gunk" at the bottom of the vat. If the vat is world-sized and the time span geological, we should begin to see floating globs of protein, called coazervates, and eventually membranes, giving rise to individuals and individual histories. If, under some histories, the globs or membranes undergo fission, reproduction takes place and natural selection begins to operate. When there are many reproducing individuals, coevolution puts a premium on interactions between individuals—predator and prey, symbiosis, and the like. Individual-to-individual interaction, in turn, puts emphasis on signals to mediate the interactions. Ever-increasing complexity follows, stemming directly from the evolving boundaries and signals. Is this increasing complexity of coevolving signals and boundaries typical of other complex adaptive systems?

These three vignettes have signal/boundary (s/b) themes in common, but it is not easy to weave the themes into an overarching framework. Still, general answers to the questions posed depend on melding these s/b themes into a common framework. The next section makes the case that systems built around signal/boundary interactions are so widespread that much can be gained from the broader understanding provided thereby.

1.2 The prevalence of signal/boundary interactions

At a fundamental level, thought itself depends on our penchant for seeing the world in terms of the bounded shapes we call *objects*. To see an object, we must organize the perpetually novel, confusing array of light signals striking the eye into familiar, bounded shapes. Once we organize our sensory input into objects, we can go on to parse the world about us in increasingly sophisticated ways. Present-day human society depends on cataloguing boundaries that range from geographic, linguistic, and political boundaries down to the boundaries that define bodily organs and the membranes that compartmentalize biological cells. In each case there are matching signals that facilitate interaction and control.

Among the systems that depend on signal/boundary interactions are several that now play important roles in life—the Internet, cancer, equities markets, ecosystems, and language acquisition, to mention a few. A short list of questions about these systems follows. Each question is accompanied by a suggestion of a formal approach that would help answer the question. These approaches will be investigated later.

What new functions will the Internet offer as bandwidth and mobility increase dramatically?

In the early 1980s few would have predicted the pervasive influence of the Internet that has resulted from applications such as Google, YouTube, and Second Life. What does the Internet have in store for us now? Though this question is challenging, it is only one example of broader questions about networks. Networks can be used formally to describe a wide range of systems, including the evolution of species, Darwin's "tree of life" (Allen 1977), the structures of nervous systems (Rochester et al. 1955), and the complex feedback loops within a biological cell (Alberts et al. 2007). In these descriptions, the network's nodes represent bounded entities (species, neurons, organelles) and the connections between nodes represent the flow of signals between the entities. There are formal methods for studying and comparing networks (Newman, Barabasi, and Watts 2006), but so far there has been less concern about the coevolution of parts of networks. The mechanisms of coevolution will be a major concern of this book.

How can "designed" signaling proteins be used to interrupt or modify undesirable changes in cell activity, such as stages in the progression from normal cells to cancerous ones?

Here the emphasis is on the filtering aspect of boundaries. External signals can *induce* cells to modify their internal signaling networks. Cells can also evolve boundaries that make them *competent* to resist these external signals. The interplay of induction and competence is the essence of developmental biology

(Buss 1987). There is good reason to believe that designed proteins can modify competence and induction to enhance the immune system's ability to respond to complicated infections such as bird flu and HIV.

In view of the distributed nature of markets, are there activities open to individual agents that will damp violent swings?

The most interesting signal/boundary systems are composed of *agents*—bounded subsystems capable of interior processing. In many cases, these agents are organized hierarchically, with an executive at the top. However, there are other systems, including bird flocks, fish swarms, and markets, that have no central executive. In such systems, the behavior of the whole depends on local interactions, a phenomenon called *distributed control* (Han, Li, and Guo 2006). Adam Smith (1776) used the metaphor of the invisible hand to describe distributed control in markets, but the mechanisms that give rise to the invisible hand are still only partly understood. Now we face the broader question of how to encourage innovation in market economies.

Why do tropical rainforests exhibit exquisite ecological innovations and great diversity even though they grow on the world's poorest soils? How can we preserve these ecosystems?

In a rainforest, you may walk a hundred paces or more before you encounter two trees of the same species. And a single tree in a tropical rainforest may host more than a thousand distinct species of insects. How do these distinct species manage to coexist on a single tree? Like a rainforest, a contemporary economy exhibits a great array of products and services. We know only bits and pieces of the mechanisms that give rise to and sustain this diversity. With a decent framework, we would be able to build knowledge of common underlying mechanisms from insights obtained in special cases.

What enables an infant to acquire the complicated human signaling ability called language? Are there ways of improving language acquisition?

Language is, above all, a social phenomenon mediating interaction between human agents. Within a few years, a child becomes a sophisticated user of language, able to express desires and discuss complicated situations not available to immediate sensory input. There is still no consensus on the broad outline of this process, let alone on the best methods for enhancing language acquisition. On a broader scale, we confront the role of learning in *autonomous systems*—systems not driven by current stimulus alone. Any "theory of mind" must include this notion of autonomy.

It is, of course, possible to investigate these questions one by one, using ad hoc techniques. But that approach sets aside the possibility that answers to one question may help in answering the other questions. The signal/boundary components in each case, along with other similarities, suggest the possibility of a unifying framework. Such a framework offers two important advantages:

- It lets us transfer discoveries from a well-studied discipline to other disciplines that are less accessible or less familiar.
- It tells us where to look for relevant facts or mechanisms when confronted with new or unfamiliar consequences of signal/boundary interactions.

The "where to look" guidance of a formal theory is often overlooked. Consider that paragon of unifying theories, Newton's theory of gravity. Even now, after hundreds of years, Newton's theory plays an important role in determining the trajectories of interplanetary probes, an application not even vaguely imagined in Newton's time. In a similar vein, Maxwell's theory of electromagnetism, which enables us to control electromagnetic signals, led to the construction of digital computers. And Darwin's theory of evolution led to the use of genetics to study the effects of new diseases. In each case, without the guidance of theory we would have been severely handicapped.

Developing a formal theory is akin to learning the rules of a game. For example, Newton's laws give us a set of rules for the movement of objects. Once the rules are known, it becomes possible to make principled predictions and choices, such as directing movement toward a desired outcome. The quest for "rules of the game" that apply to the broad spectrum of signal/boundary systems is the concern of this book. As we have already seen, signal/boundary systems evolve continually. So the "rules" must provide for origins of new signals and boundaries and their selective modification. For example, we cannot really understand the human body, or indeed multi-celled organisms in general, without understanding the origins of their organization. The human body is more like an ecosystem than one might at first imagine. Not only is there a great diversity of body cells, but for each body cell there are roughly 100 bacterial cells co-existing in the body at the same time. These co-existing bacterial cells are mostly beneficial, or at least harmless, but we know little of their individual roles. To understand those roles we must understand the coevolution of the bacteria and the body cells. The theories of population genetics and foodwebs can help us in this task, but we need a broader framework to take those theories beyond the specifics they examine.

To get a better idea of the form of a theory based on signal/boundary rules and mechanisms, it is helpful to take a close look at a quintessential signal/boundary system: a tropical rainforest. Even defining a rainforest requires close attention to its signal/boundary interactions.

1.3 Tropical rainforests: an important example

Tropical rainforests, which have fascinated humans ever since the age of exploration, still hold many mysteries. Even simple interventions in a rainforest have complicated and long-lasting effects. For example, extensive slash-and-burn agriculture in rainforests has intricate effects on the global climate that are only partly understood. Whole books have been written about the complex interactions in rainforests (e.g., Forsyth and Mivata 1984). Here I will confine myself to features that highlight signal/boundary interactions.

I should emphasize again that this book is not primarily about rainforests, or even biology; its objective is to better understand the origins and effects of signal/boundary interactions, wherever they occur. To this end, it is helpful to group rainforest interactions into four categories that are relevant to all signal/boundary systems: *diversity*, *recirculation*, *niche*, and *coevolution*. I'll now discuss the four categories, ending each discussion with a brief look at the role of the category in other signal/boundary systems. In later chapters, I'll look at these correspondences more carefully.

Diversity

The diversity of flora and fauna in a tropical rainforest is striking, especially for those who have experienced only temperate-zone forests. The mystery of this great diversity is compounded by the fact that a rainforest grows on impoverished soil—the heavy rains leach all the nutrients from the soil, cascading them into the nearest stream. Why doesn't this impoverishment result in a forest in which a few species struggle for survival, as in a desert? What mechanisms make it possible for so many distinct species to coexist?

The organisms in a rainforest act much like a stack of catch basins with overflow spouts. Each organism uses resources passed on from other organisms. In simple cases, a predator uses resources acquired in eating prey. In a rainforest there are much subtler interactions. For example, the resources acquired by one organism may be processed by a succession of organisms, much as happens with the cells in the body of a multi-celled organism, or organisms can exchange surplus resources gathered elsewhere in the forest. Resources can even be used over and over again via *recirculation,* a mechanism that is important in sustaining diversity in impoverished circumstances.

In the modern world, diversity is closely related to specialization. Production lines replace individual craftsmen with specialists acting in sequence; arrays of interacting mechanisms in a wristwatch or an airplane depend on carefully defined signal/boundary interactions; languages and their dialects are notoriously diverse. Are there common mechanisms that generate diversity in different signal/boundary systems?

Recirculation

The diverse organisms in a rainforest interact in a finely tuned way to keep nutrients from being lost to the leaching of the forest floor. As one example, consider a typical tree-dwelling bromeliad, a plant belonging to the same family as the pineapple. The bromeliad's leaves form a basin that holds water. That basin acts as a sanctuary wherein various insects and frogs lay their eggs. The waste products of the growing larvae, in turn, provide nutrients that are useful to the bromeliad. The bromeliad becomes an ecosystem in itself, much like a spring pond in the temperate zone. It is easy to multiply these examples of recirculation. Ants cut leaves and store them as the base for an underground fungus garden. A sloth becomes part of a tree, its fur serving as a home for algae, and when the sloth occasionally descends to the ground it plants its hard little feces at the base of its tree, so that they fertilize its home. And on and on.

Even the lowly slime mold manages to conserve resources. Slime molds, ubiquitous in rainforests, use signals to accomplish sophisticated "social" interactions that enable them to save resources in hard times. Under normal conditions, molds of a given species are mobile single cells, moving about like amoebae. However, when resources become scarce, the individuals of that species emit a chemical signal that causes them to aggregate. Once aggregated, the individuals differentiate into a base and a stalk that supports a capsule containing hard, inert spore cells. All the cells then die except the spore cells; the spore cells lay quiescent until times get better, a clever way of reducing resource expenditure during a period of scarcity.

The diversity of highly specialized species plays an important part in the conservation of resources by recirculation. Boundaries, both implicit and explicit, confine the nutrients so that they

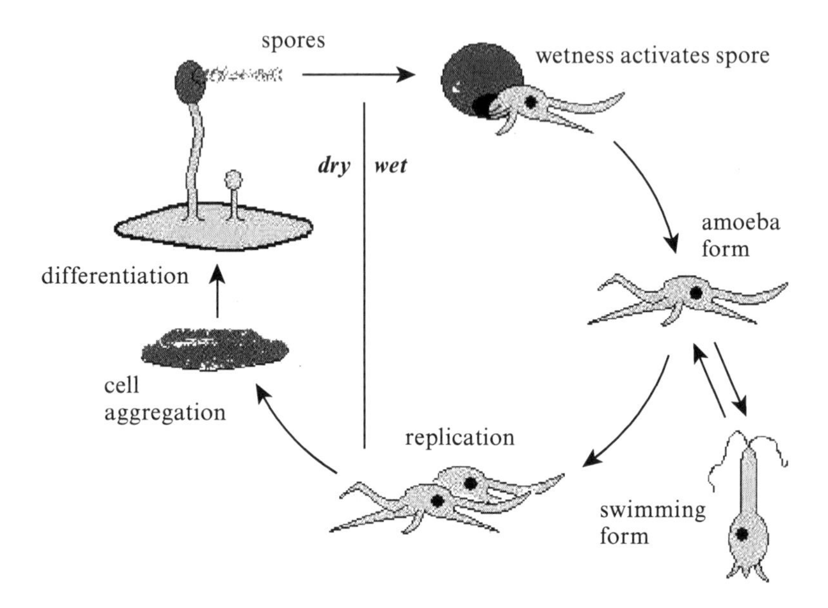

Figure 1.2
Slime mold.

can be picked up by other organisms. The resulting possibilities for recirculation induce a "multiplier effect" analogous to the "multiplier effect" of economics: In a money-based system, if I purchase the services of a carpenter, the money doesn't "disappear" when I pay the carpenter. Rather, the carpenter uses it in turn to make purchases of lumber, food, and other things, perhaps retaining some in savings. The lumber retailer, in turn, pays a wholesaler, and so on. A dollar of new purchase can thus have the effect of many dollars when one looks at the system as a whole. So it is with the nutrients in the rainforest: they pass from organism to organism.

As with diversity, recirculation is a common feature of systems exhibiting signal/boundary interactions. The multiplier effect in an economy has already been mentioned. Different kinds of feedback made possible by recycling have major effects on metabolic networks (Alberts et al. 2007). The continual cycling of pulses through entangled loops of the central nervous system is a critical feature of the behavioral flexibility of mammals. The cyclic adaptations of the flu virus allow its recirculation, sometimes at epidemic levels, through its host populations. We can hope to learn more about all these phenomena if we can achieve a better understanding of the networks describing signal/boundary interactions.

Niche and hierarchy

In Peru, small patches of rainforest exhibit the great variety of species typical of rainforests everywhere, but occasionally one encounters a lawn-sized patch almost entirely occupied by a single species of tree: *Duroia hirsute*. According to local legend, such "devil's gardens" are cultivated by an evil forest spirit. Until quite recently there were only conjectures about a natural explanation for these patches. However, there is a species of ant (*Myrmelachista schumanni*) that lives in small cavities in the monospecific trees occupying the "garden." We now know that the ants "weed" this "garden" by injecting formic acid into the seedlings of other plant species, much as if they were attacking another insect (Holldobler and Wilson 1990). The ants probably attack all seedlings that don't emit the scent of the favored tree. This signal detection is similar to the action of an immune system, which destroys all cells that don't send the "self" signal. By setting up this bounded niche, the ants create a locally ideal environment.

"Niche" is one of the most important signal/boundary concepts and, at the same time, one of the least understood. Even in ecology, where where the niche concept is ubiquitous, it is well defined only in special cases. The combination of niches and diversity gives the rainforest an intricate patchiness that makes it difficult to extract cause and effect. How do different organisms find the appropriate organisms (prey, symbionts, mates) with which to interact? The broad answer is that signals (scents, sounds, visual stimuli, and so on) offer the possibility of something more than random interaction. Indeed, the rich sights, sounds, and scents of the tropical forest make an immediate and indelible impression on a visitor. This enabling role of signals raises a deeper question: How do the signals that make finely tuned niches possible come into being?

Because there are niches within niches, web-like hierarchies result. In signal/boundary terms, there is a hierarchy of enclosing boundaries, with matching signals at each level. The bromeliad just discussed is a clear example, providing a boundary within which further bounded agents (insects and frogs) exist. Each insect and each frog is, in turn, a hierarchy of enclosures, starting with the internal organs and ranging down to the organelles within individual cells. Each of these boundaries is semi-permeable, permitting easy passage for some resources and difficult passage for others. This semi-permeability encourages local concentrations of resources and reactions that enhance recirculation. The resulting niches offer new possibilities for interaction and increasing diversity. How are these possibilities realized? Part of the answer resides in the *coevolution* of the denizens of the forest, our next topic.

Niches and hierarchies are common to all signal/boundary systems. Here are a few hierarchies: The biosphere can be divided into geographical regions (e.g., tropical rainforest, temperate-zone woods), which in turn consist of niches (e.g., bromeliads, spring ponds), which are further divided into populations (e.g., frogs, snails, dragonflies, algae), and ultimately individual organisms. Multi-celled animals can be progressively divided into organs (e.g., heart, brain), organ components (e.g., heart muscle, prefrontal cortex), cells, and organelles. The global economy can divided into national economics, then into industries or markets (e.g., coal, steel), firms, and production steps. In real signal/boundary systems, hierarchical arrangements are more intricate than these simple sketches indicate, with many web-like features.

Because of the enclosure-within-enclosure-within-enclosure form of these niches, it is necessary to pick a "lowest level" at which to lay out a corresponding network. The lowest level is typically chosen as a set of easily distinguishable components relevant to a question of interest. For example, in studying the global economy, we could choose industries as the lowest-level agents. Of course, these "lowest-level" components can often be further decomposed into still lower-level components, proceeding step by step to the level of atoms or below. But inappropriate levels of decomposition can make it difficult to answer the question of interest. Current studies of networks (Newman, Barabasi, and Watts 2006) using notions of community and synchrony within subgroups help to make the niche concept more precise. However, it is noteworthy that few network studies concentrate on the *formation of boundaries* within a network. And there is even less study of *mechanisms for the formation of hierarchies*—mechanisms that would explain the pervasiveness of hierarchies in natural systems. That is due in part to the extreme difficulty of the mathematics of such processes; however, it is also due in part to the current focus of network studies, which are not mechanism-oriented. Still, studying the formation of boundaries within networks offers a broad approach to all signal/boundary interactions.

Coevolution

Sometimes natural selection's status as a "real" scientific theory is questioned on the ground that it produces no predictions. However, in Darwin's papers (see Allen 1977) one can see a quite singular prediction stemming from his knowledge of the "fine-tuning" produced by coevolution. Darwin wrote extensively about orchids. The Madagascan orchid (*Angraecum sesquipedale*), sometimes called the "comet orchid," fascinated him because it has a whip-like nectary more than a foot long. Of what use could such a nectary be? A pollinator would need a foot-long "tongue" to take advantage of the nectar. Darwin, with his extensive knowledge of orchids, knew that similar orchids were pollinated by moths, but there was no known Madagascan moth with a foot-long proboscis. Darwin boldly predicted that there must be such a moth in Madagascar. Just recently, the predicted moth, a sphinx moth, was filmed for the first time.

Some sphinx moths mimic hummingbirds almost perfectly, having a long proboscis that mimics a hummingbird's beak and transparent wings that beat rapidly, producing a hummingbird's "angry bee" sound. Most moth predators stay well away from a sphinx moth. Such mimicry is a prevalent outcome of coevolution. Orchids too are mimics. For example, there are orchids, such as the "bee orchid," that mimic particular insects so well that the insects attempt copulation with them, becoming pollen carriers in the process. Note that each insect-mimicking orchid species requires a particular insect for pollination—an extreme form of coevolution. The process of mimicry is an important adaptive signal/boundary interaction.

In coevolution, each new species offers new possibilities for interaction. These interactions can become increasingly specialized, as in the case of the comet orchid and the sphinx moth, offering still further possibilities for interaction. Combining this progressive specialization with the "multiplier effect" mentioned above in the subsection on recirculation gives a tantalizing comparison between the richness of ecosystems and the richness of factory systems. Again boundaries and enclosures are crucial. Specialization requires parsing the original whole (e.g., the individual craftsman, a generalist) into stages (e.g., a sequence of agents forming a production line, the specialists). In the rainforest, we have interlocking networks of specialists (symbionts, mimics, parasites, and so on), greatly enriched by hierarchical enclosures and great diversity. Exchange across these interlocking boundaries, with local enrichment and specializations, allows for complicated "multiplier effects" (Krugman et al. 2010, Samuelson and Nordhaus 2009).

As has already been mentioned, the progressive specialization of interactions has an analog in economics. As an illustration, consider Adam Smith's famous example of the pin factory in *The Wealth of Nations* (Smith 1776). Ever since their invention, metal straight pins had been produced by craftsmen in a series of tedious operations. As a result, they were expensive. Around the time Smith was writing, the process began to be shared out among a group of specialists, in effect creating a production line. The price of straight pins went down by a factor of 10.

I have also mentioned the human body as an example of coevolution. It is a good thing that bacterial cells weigh so much less than body cells; otherwise we would be carrying hundreds of pounds of extra weight. The intensity of natural selection forces the co-existing bacterial cells into an "ecosystem" of coevolved interactions quite similar to the diverse interactions in a tropical rainforest. The resulting coevolved specializations must offer advantages to the whole multi-celled organism—we can even describe the advantages in some cases—but the larger picture is difficult to see.

Every time we examine a signal/boundary system closely, we see coevolution as a pervasive feature. Products of coevolution occur in all aspects of human endeavor, from the sciences to the arts, including such subtleties as imitation and mimicry. Diverse languages coevolve through borrowing and translation. Global trade depends directly on local specialization and global coevolution. The coevolution of crops and bacteria has constituted a major problem ever since the earliest days of agriculture. In sports, we see the rules coevolving with the capacities of the players. In each case, important signal/boundary interactions are involved.

Summary

This look at tropical rainforest interactions, brief though it is, offers guidance in our attempt to understand signal/boundary interactions in general. The mechanisms and interactions falling under the broad categories *diversity, recirculation, niche and hierarchy*, and *coevolution* are central to understanding a wide range of large-scale and small-scale signal/boundary systems. If we can fit these categories within an overarching framework, it will further our quest for a general understanding of signal/boundary phenomena.

1.4 Mechanisms

In the examples considered so far, signals and boundaries are made up of a multiplicity of interacting parts, ranging from the molecules that define membranes to the utterances that are sequenced to yield spoken language. This common grounding in interacting parts, combined with the categories of interaction just examined—diversity, recirculation, niche, and coevolution—leads us directly to a rapidly growing field of study: the study of *complex adaptive systems* (abbreviated *cas*, which may be read as either singular or plural as context requires). The components of a *cas* are bounded subsystems (*agents*) that adapt or learn when they interact. Markets, languages, and biological cells all fit the *cas* framework, the agents being, respectively, buyers and sellers, speakers, and proteins. (See Mitchell 2009.)

The data sets for the three complex adaptive systems just mentioned—markets, language, and biological cells—are enormous. Unfortunately for present purposes, there is a shortage of integrative information—that is, information about the mechanisms that generate these data sets. For example, to combat a disease that has seasonal variations, such as flu, we must discover the mechanisms that provide the seasonal variations. Currently, because we know some of the mechanisms underlying flu variation, we can begin to "tune" the flu vaccine each year, giving the recipients immunity to the most likely variants. But when we don't know the mechanisms, it is as if we were observing a game without knowing the rules. When common mechanisms can be extracted, it is also easier to understand the idiosyncratic features of particular signal/boundary systems.

Much of this book is about the quest for such mechanisms, and the rules that govern them.

Owing to the current lack of a full-blown formal theory of *cas* mechanisms, the *cas* framework cannot serve as a ready-made candidate for an overarching framework of signal/boundary systems. *Cas* concepts can nevertheless be helpful. Studies of particular complex adaptive systems, as we will see, have developed new tools for examining diversity, recirculation, niche, and coevolution. Because these topics figure in *cas* studies, the quest for an overarching theory of signal/boundary systems should contribute to *cas* theory and vice versa. Still, the lack of an overarching theory means that this book must often proceed in exploratory mode. That is not a bad thing—both as humans and as scientists we generally understand much more than we can establish through logical argument. There are a variety of tools we bring to bear: experiment, calculation, modeling, comparison, everyday reasoning, analogy, metaphor, even taste.

Though studies of the dynamics of complex adaptive systems and of signal/boundary systems are still in their early stages, the dynamics of mechanical systems have been studied for a long time. It will pay to look at this expertise to see what a similar mastery of signal/boundary dynamics would require. Aerodynamics provides a useful example. Consider the steps taken to extract the critical elements that influence the flight path of an airplane. Here we have another good example of the importance of questions and goals in determining centrality and detail. It was part of the genius of the Wright brothers to distinguish between the dynamics of bird flight and the dynamics of an airplane. (See Goldberg 2002.) Before them, those attempting artificial flight had tried to imitate too many elements of bird flight. Even today, the use of wing flapping for both propulsion and control poses a challenge to aerodynamic theory because there are many nonlinear interactions in a flapping wing. When applied to turbulent flow, Newton's equations soon become overwhelmed with details. The Wrights succeeded by recasting the problem so as to separate propulsion from flight control, thereby clearing away many of the subtle details resulting from the complex interactions. By concentrating on the dynamics of these separate elements of mechanical flight—air flow across airfoils, mechanical controls for attitude, propellers, and so on—they were able to produce a controllable airplane. It will pay us to take a similar approach to signal/boundary dynamics, concentrating on the mechanisms that generate the dynamics.

The interactions singled out in section 1.3—diversity, recirculation, niche and hierarchy, and coevolution—play roles similar to those of air flow, mechanical controls, and propellers in the Wright brothers' decomposition. The problem is to relate these elements to overall questions about the origin and the evolution of signal/boundary interactions. Values for these critical signal/boundary elements are analogous to readings from an airplane's control panel—they determine the current state of the components that are responsible for the system's dynamics. Knowing these "readings" gives us the opportunity of knowing what will happen next. Before we can have a signal/boundary theory, we must first determine how these elements can serve as grist for adaptation; then we must examine how adaptive mechanisms provide for the coevolution of signals and membranes.

1.5 Unification

There are two broad questions that apply to the full range of signal/boundary systems:

• What are typical steps in the formation and evolution of the complex signal/boundary interactions involving "generalists" (e.g., craftsmen) and "specialists" (e.g., workers on a production line)?

• What mechanisms generate these signal/boundary interactions?

Answers to these questions come more easily when different signal/boundary systems can be compared in a general framework. Relevant data that are available or easily acquired in one signal/boundary system can then be used to guide understanding of other signal/boundary systems in which similar data are absent or difficult to obtain.

The *cas* framework provides a first step toward a general framework by offering a general way for defining *adaptive agents*. Adaptive agents are defined by an enclosing boundary that accepts some signals and ignores others, a "program" inside the boundary for processing and sending signals, and mechanisms for changing (adapting) this program in response to the agent's accumulating experience. Once the signal/boundary agents have been defined, they must be *situated* to allow for positioning of the relevant signals and boundaries. That is, the agents must be placed in a geometry that positions populations of agents of various kinds and localizes non-agent "resources." Within this geometry, agents must be able to form *conglomerates* that yield higher levels of organization with new boundaries so that the framework can capture hierarchical organization.

The foregoing desiderata place strong requirements on an overarching framework—requirements that, taken together, are not easy to meet. Chapter by chapter, this book introduces a sequence of concepts that can be melded into an appropriate framework:

classifier systems for defining signal-processing programs

tags for directing signals

tagged urns for defining semi-permeable boundaries

genetic algorithms to provide for the adaptation and coevolution of agents

dynamic generated systems that provide a "grammar" for the framework and bring mathematical tools to bear.

The book's major task is to fit these concepts together to form a well-defined framework.

To keep sight of the overall goal, it is useful to fit the concepts together, step by step, as they are introduced. To this end, each chapter illustrates the new concepts by using biological cells as a running example. Biological cells, with their sharply delineated set of signals and boundaries, offer an easier way to tie the concepts together than the more diffuse signal/boundary complexes of rainforests. In addition, a biological cell's well-described metabolic network, regulated by a combination of signals and boundaries, illustrates the persistence of signal/boundary hierarchies over time scales ranging from milliseconds to millennia. The cell's many compartments and structures (mitochondria, ribosomes, lysosomes, the Golgi apparatus, etc.) use signals—proteins (with fanciful names like "hedgehog'), promoters (repressors) that turn genes on (off), scaffold proteins, hormones, and so on—to establish and control the network. (See Alberts et al. 2007, p. 321; Felicity et al. 2010; Good, Zalatan, and Lim 2011.) The running example describes the relevant cell structures and then applies the newly introduced concepts to those structures. In particular, the running example shows how the concepts and structures relate to the four basic characteristics extracted from the rainforest example: diversity, recirculation, niche, and coevolution. Even at this early stage it is evident that the cell's meta-

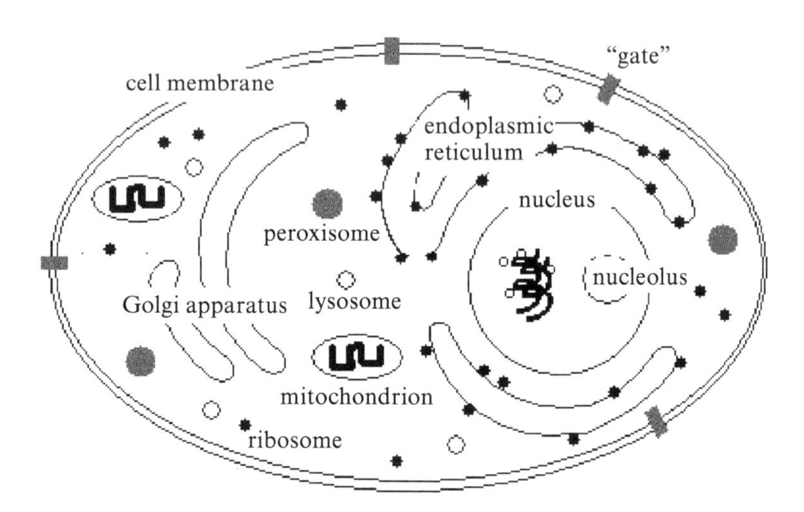

Figure 1.3
A schematic diagram of a cell.

bolic network depends on extensive feedback (recirculation) to keep it within workable limits, that the variety of cell structures (diversity and niche) offer ways of efficiently employing incoming resources to steer the cell through "shocks," and that the coevolution of signals and boundaries in the cell produces a progression of cell types, from the simplest self-reproducing single-celled protozoa to the complicated cells found in multicellular eukaryotic organisms. (See Alberts et al. 2007.) The running example uses the concepts as they are introduced to build these characteristics into the overarching framework.

1.6 What is to come

A series of simple models that fit together to provide an integrated overview of signal/boundary systems will be presented. Here are some highlights.

Theory

Models of agents, networks, and adaptation motivate a closer examination of theory's role in the study of signal/boundary systems. Chapter 2 discusses tactics in theory construction and the results that can be expected from a relevant signal/boundary theory. As might be expected, this examination suggests a theory that provides direct representation and manipulation of boundaries. At the same time, the theory should connect well with extant theories, allowing as much use of available mathematics as is possible. One way to meet these desiderata is to adapt the urn models of probability theory discussed in chapter 6 to the study of the movement of signals through boundaries.

Signal-processing agents: reactions and interactions

Chapter 3, the starting point for the discussion of signal processing, deals with agent-based models. Agents are omnipresent in signal/boundary systems. For the Internet, sites act as agents processing messages; for the immune system, antibodies act as agents counteracting invading cells (antigens) through a complicated exchange of signals (proteins); for markets, the agents are buyers and sellers and the signals are bids; for ecosystems (e.g., rainforests), the agents are organisms interacting via complicated resource exchanges; for biological cells, organelles serve as agents interacting via exchanges of proteins; for language, the agents are humans communicating by sequenced utterances; for the central nervous system, agents are clusters of neurons that interact via electrical pulses sent over branching interconnections (axons).

An agent is defined by an "outer" (containing) boundary (e.g., the outer membrane of a biological cell) that determines what kinds of signals can affect its activities. This outer boundary usually contains a substantial fragment of the agent's signal-processing apparatus. Because agents in a complex adaptive system can aggregate to act as agents at a higher level, in effect choosing a boundary at that higher level, the choice of an agent's outer boundary determines the level of detail and the kinds of questions asked. Once an agent-defining boundary has been chosen, all interior boundaries are treated as parts of the agent's signal-processing apparatus; signals interacting with the interior boundaries coordinate the agent's interior workings.

In chapter 3 we will see that, regardless of the details of the agent's interior apparatus, a *cas* agent can always be defined in terms of a set of signal-processing rules called *classifier rules.* Each rule accepts certain signals as inputs (specified by the *condition* part of the rule) and then processes the signals to produce outgoing signals (the *action* part of the rule). Formally, both signals and boundaries can be defined using strings of "letters" drawn from one basic "alphabet." This limited alphabet has a counterpart in cellular biology, where both the structure of protein signals and important parts of "gateways" in semi-permeable membranes are defined using an alphabet of twenty amino acids.

Because signal-processing rules can exchange signals in closed sequences, we can construct loops (subroutines) that can execute any computable manipulation of strings. That is, anything that can be programmed for a computer can be executed by appropriate arrangement of these signal-processing rules. This *computation universality* ensures that a framework based on classifier-defined signal-processing rules is powerful enough to treat any signal/boundary system.

Chapter 3 also shows that interlocking chemical reactions can be described as rule-based systems, with the added constraint that elements defining the signal strings are conserved. That is, the rule's outputs must be formed by recombining the letters present in the rule's inputs. Though the conservation-of-elements requirement is a departure from the usual conventions in computation, rule-based agents satisfying this constraint are capable of carrying out any computation when supplied with sufficient elements, . The conservation requirement opens up the possibility of studying signal/boundary interactions by exploiting *particle mechanics* (less formally, billiard-ball models), a well-developed part of classical physics.

The basic idea of particle mechanics is that interactions are handled as random elastic collisions between individuals (e.g., atoms in a well-mixed gas). In this model, signals are treated as particles (billiard balls) and the broadcast of signals is handled by random diffusion. Different signals are identified with different "colors" of billiard balls, and reactions occur when billiard balls collide. When a reaction takes place, the colors of the colliding balls are changed to the colors of the products of the reaction. This relation between reactions in a billiard-ball mechanics and signal processing in an agent is examined in section 3.4.

Signal-processing networks

Networks (more formally, graphs) are used in chapter 4 to move from the interactions implicitly defined by signal-processing rules to an explicit representation of those interactions. Each *node* in the network represents a rule in the classifier system. *Edges* connecting nodes in the network represent interactions between the rules. Two nodes are connected by a *directed edge* when the output of the rule labeling the first node is accepted by a condition of the rule labeling the second node. Networks in a typical signal/boundary system involve many loops and cycles because of the prevalence of recirculation. These loops and cycles impose analytical difficulties not present in the tree-like networks commonly used to describe foodwebs or cascades of protein interactions. As an additional complication, most signal/boundary interactions are *situated*—that is, confined to sites, such as different geographic locations in an ecosystem. In a rule-based format, this constraint results in clusters of rules operating at different sites. In a network format, nodes belonging to different sites have few edges connecting them; signals move from site to site only through nodes representing boundary (entry/exit) rules. Interestingly, these analytic difficulties suggest an approach centered on the loops, using them as building blocks that are combined and recombined during evolution.

Adaptation, recombination, and reproduction

With the foregoing complications in mind, the same fundamental question arises in each of the examples introduced above: How do finely tuned boundaries and signals come into being?

It is clear that coevolution, in one sense or another, is involved in every case. Indeed, this book's major thesis is that the key to understanding complex signal/boundary interactions is an understanding of their progressive coevolution.

Chapter 4 discusses the analysis of coevolving agents, and their continually changing networks of interaction, by extracting "building blocks" that serve as grist for adaptive mechanisms. In networks, building blocks are pieces (e.g., communities) that can be reused in many different contexts. For example, in metabolic networks, the Krebs (citric acid) cycle, consisting of eight protein catalysts, is a building block found in every aerobic organism, from carrots to elephants; it serves as a kind of fundamental "Lego block" from which many complicated metabolic networks can be constructed. (See Smith and Morowitz 2004.)

Chapters 5 and 6 discuss the processes of adaptation and the adaptive mechanisms that underpin evolution and coevolution. Mutation and crossover are two of the primary adaptive mechanisms underlying evolution and coevolution (though there are other mechanisms, among them intrachromosomal duplication). With crossover the emphasis is on generating new signals and boundaries by "cross-breeding" existing signals and boundaries. Successful cross-breeding typically relies on recombination of tested building blocks such as the Krebs cycle.

Specialization is a typical outcome of coevolution and cross-breeding. As the system adapts, inefficient single-stage processes are replaced by sequences of more specialized processes. Adam Smith's example of the pin factory illustrates the replacement of a one-stage process with a multiple-stage process. A multi-skilled craftsman is displaced by a sequence of specialist agents, each specialist adapting and refining some particular skill of the generalist (such as drawing wire or adding a head to the pin). For the process to work, each specialist must receive input from a predecessor agent and pass the processed input on to a successor. The pin factory, though a special case, illustrates three factors in signal/boundary interactions:

New boundaries distinguish the specialists.

Resources flow from specialist to specialist.

Signals synchronize interactions.

It is instructive that nearly 300 years after Adam Smith highlighted this adaptive transformation we still lack a general theory of the origin of production lines. What is difficult in Smith's example must be at least as difficult in the general case. Progress depends on finding adaptive mechanisms, such as cross-breeding, that make specialization possible.

Chapter 6 points up the simplicity of cross-breeding when strings are used to define rules. A segment of one string is exchanged with a corresponding segment of another string. Moreover, when signals are also strings, selected short segments—tags, such as "headers" for messages on the Internet, common "motifs" in proteins, "promoters" in chromosomes, and so on—come to play a special role in adaptation. A rule condition can then select signals according to the tags they carry; the tag serves as an "address" to guide the signal to the rule. The widespread appearance of tags in signal/boundary systems suggests that tags serve well as building blocks that can be recombined to generate new signals and rules.

Urn models and multi-armed bandits

Models of agents, networks, and adaptation motivate a closer examination of theory's role in studying signal/boundary systems. One way to bring mathematics into play is to adapt the urn models of elementary probability theory to the study the movement of signals through boundaries.

Chapters 7 and 8 lay out the use of tagged-urn models to represent boundary hierarchies, in the process relating these models to the models discussed in earlier chapters. Some of the earliest models of probability concern the drawing of balls from an urn filled with different proportions of black and white balls. (Recall the "blackballing" process used in classical Athenian democracy.) By moving balls between multiple urns, we get models of diffusion. By using balls of different colors and assigning conditions under which these balls are allowed to enter or exit urns, we get models of different signals moving through semi-permeable boundaries. Chapters 9 and 10 examine the use of tagged-urn models, and related models, as facilitators for a theory of niches, language, and grammar.

An overarching framework

Chapter 11 introduces and explains *finitely generated systems* as a way of tying the models examined earlier into a single framework. Chapters 12–14 examine this framework in detail, using a simple finitely generated version of the development of a multi-celled organism from a "spore" to illustrate the framework. Chapter 15 provides an exposition of Markov processes as a way of deriving theorems that apply to full range of systems that fit within this overarching signal/boundary framework. The book closes with a compressed review of the path laid out—it is a path that leads to a more comprehensive understanding of signal/boundary systems, but so far it is a path, not a highway.

2 Theory and Models: General Principles

For signal/boundary systems there is almost always a large gap between the data available and an explanation of the interactions that generate that data. This gap, currently quite large for s/b systems, is familiar in all the sciences. For millennia, humans observed, and speculated about, "laws" governing moving inanimate objects—objects ranging from falling stones and flying arrows to planets. Fallacious "laws" were common: "a moving object always comes to rest," "heavier objects fall faster," and so on. All the while, the list of observations and examples was growing longer. It took Newton's simple laws to bring this diverse array of observations within a common framework that both predicted what would happen in unfamiliar situations and made control possible. The framework also fostered new methods of teaching (e.g., general principles of fluid flow) and new artifacts (e.g., turbines). Our objective now is to examine the role that theory can play in the study of s/b systems.

2.1 Why theory?

When we face difficult scientific questions that cannot be answered by inspection or by trial and error, formal theory has an essential role. Consider the question posed in chapter 1 about the origins of language: Did the rich network of signals and boundaries at the edge of the African savannah stimulate rapid changes in the primate brain, leading to the emergence of language? An answer to this question would bear directly on our understanding of humans and human organizations. But it remains a conjecture. Why? The simple answer is that we have no coherent theory of emergent phenomena in either complex adaptive systems or signal/boundary systems. Without theory, we lack the critical guide that would tell us where to look for the data that will prove or disprove the conjecture.

Common usage often treats "theory" as synonymous with "speculation," but scientific theory has a different role. When a question is approached scientifically, the assumptions (premises) must be made explicit. Moreover, the assumptions must be formalized so that standard rules of deduction can be used to derive answers. These rules constrain the argument to simple obvious manipulations, much like moving pieces according to the rules of a game. Euclidean geometry, originally used as a theory of the physical world, is a canonical example of a deductive scientific theory.

Typically, the rules of deduction are drawn from symbolic logic, in which the rules manipulate symbols without reference to the interpretation or the meaning of the symbols. That is, the manipulations are *syntactic*, depending only on the arrangement of the symbols. This makes the manipulations even more like moving pieces on a game board. Assigning meanings to the symbols has no effect on the argument itself; the assignment only affects the *interpretation* of the derived consequences. This syntactic approach comes close to being a sine qua non for theoretical science. Matters of speculation and interpretation are moved from the argument back to the premises. That makes the conclusions immune to the often misleading biases of rhetorical argument or persuasion. Interpretation and intuition may suggest a path from premises to consequences, but the actual derivation in a formal theory is a syntactic, non-interpretive process.

2.2 First steps

The first step toward a signal/boundary theory, then, is to phrase the questions we would like to address in ways that suggest premises for a deductive system. Note that different questions can lead to different theories, even when the same data are involved. Phrasing the questions and the premises is a matter of insight and taste, with close attention to parsimony.

Questions about agents provide a good starting point: How do agents arise? How do agents specialize? How do agents aggregate into hierarchical organizations? Beneath these questions is a basic question set by the fundamental *cas* agent activity of continually exchanging signals and resources: How do agents form and redirect flows of signals and resources? The examples drawn from the rainforest emphasize flows, categorizing them in terms of their effects on diversity, on recirculation, on niche and hierarchy, and on coevolution. Indeed, most *cas* agents turn out to be persistent patterns imposed on flows. As was mentioned earlier, the human body turns over most resident atoms within days, and no atom resides in the body for more than a year or so. Similarly, in a great city, individuals arrive and depart daily but the overall pattern of activity persists.

In attempting to answer the questions cited above, it is important to examine the formation of agent boundaries, both internal and external, and the effects of those boundaries on the flows. This emphasis directs attention to building blocks that can combined to define boundaries. The building blocks must, of course, be based on available data. Though there are extensive data sets for most signal/boundary systems, and we can rather easily derive a great array of reliable, sophisticated statistics from such data, such statistics do not, of themselves, reveal building blocks or mechanisms. Anatol Rapoport, one of the founders of mathematical biology, long ago pointed out that you cannot learn the rules of chess by keeping only the statistics of observed moves (Rapoport 1960). We confront the same difficulty when using statistics to study signal/boundary interactions. The interactions are just too complex (nonlinear) to allow theory to be built with the linear techniques of statistics.

However, it is often true that persistent, agent-like patterns in the data suggest building blocks. The patterns persist because they repair themselves when disturbed. That is, the patterns are analogous to the standing wave pattern in a whitewater river. To model this persistence, we must find mechanisms—generators with law-like rules for interaction—that generate and maintain the patterns. To this end, our next concern is to learn more about models based on mechanisms that generate signal/boundary patterns.

2.3 The structure of theories and models

Most models in the physical sciences use equations to define interactions mediated by forces, fields, valence, and so on. Newton's equations for gravity and Maxwell's equations for electromagnetic interaction are classic examples. The advantages of an equation-based model are clearly illustrated by Newton's equations. Before Newton, models of planetary motion were based epicycles: rotating circles, placed on rotating circles, that were adjusted until the implied motions fit the data. As the data increased in accuracy, so did the number of circles. The extensive, positional data carefully collected by Tycho Brahe (see J. R. Newman 2003) began to overwhelm the epicycle models, requiring an inordinate number of circles. Moreover, the epicycle models applied only to the special case of planetary motion in the solar system. Newton escaped this escalation by concentrating on mechanistic laws rather than direct matching of data. The result encompassed motions as apparently different as the arc of a thrown ball, the vibration of a string, the trajectory of a comet, *and* planetary motions. Newton's model has applications that weren't remotely imagined in his time, such as the calculations used to "slingshot" exploratory satellites to the outer solar system. A well-formulated equation-based model makes possible controlled experiments and sophisticated maneuvers that would not be possible otherwise.

A model of this kind, once formulated, amounts to a hypothesis about how some part of the world works. The ranges of values allowed for the variables in the model define the domain of observation to which the hypothesis applies. Sometimes these ranges are accepted without being explicitly specified. For example, it was originally assumed that Newton's equations held for all possible velocities. Now we know that Newton's equations have to be modified when objects move near the speed of light. Similarly, quantum effects modify the details of electromagnetic interactions. Accepting such limitations, we still find that the earlier theories do an amazing job of describing large chunks of the universe.

A good model or theory suggests controlled experiments to confirm or disconfirm the hypothesis it poses. More than that, models and theories have made, and do make, verifiable predictions in situations not previously encountered. A model based on mechanisms can be quite explicit in its suggestions about where to look for new possibilities. Consider the measurement of time. Even crude measurements of time have produced major changes in human cultural evolution, beginning with a determination of the times to sow and to reap. The mechanical clock combines well-known mechanisms—gears, ratchets, springs—to provide a regular motion, offering new precision in the measurement of time. This new precision made possible many of the critical measurements that underpin modern science. Indeed, there is a comprehensible series of steps leading from the discovery of simple mechanisms—levers, pumps, and so on—to the general scientific concept of mechanism. Laws framed in terms of mechanism underpin the sometimes-maligned mechanistic view of the universe. Maligned or not, the progressive description of the physical world in terms of layers of mechanisms (e.g., molecules, atoms, nucleons, quarks) has proved remarkably successful in uncovering a wide range of unsuspected possibilities, including electromagnetic waves, engines, transistors, and antibiotics.

More recently, computer-based models have provided an alternative to equation-based models. Though equation-based models and computer-based models serve similar purposes, there is a substantial difference between them. Computer-based models are a more rigorous, more elaborate version of the physicist's notion of a *thought experiment*. Both thought experiments and computer-based models provide ways of exploring the behaviors of unfamiliar mechanisms or combinations of mecha-

Level	Building Block
nucleon (proton, neutron)	quarks, gluons
atom	protons, neurons, electrons
gas or fluid	
confined (e.g., a boiler)	PVT equations, flows
free (e.g., weather)	circulation (e.g., fronts), turbulence
molecule	mass action, bonds, active sites
organelle	membranes, transport, enzymes
. . .	. . .
ecosystem	predation, symbiosis, mimicry

Figure 2.1
Building blocks used in the physical sciences.

nisms. Computer-based models have become a major tool for investigating complex adaptive systems because they can handle conditional "IF this happens, THEN take this action" interactions. Such interactions pose difficulties for traditional equation-based models because, in mathematical terms, they are nonlinear (non-additive).

Here is a brief comparison of the two kinds of models.

Equation-based models

Mathematical derivations are used to examine equation-based models, so the results are valid over the full range of values allowed for the variables. As a consequence, interactions can be "eyeballed" or intuited over a wide range of possibilities. There is, however, a severe limitation: the nonlinear, IF/THEN conditionals typical of complex adaptive systems do not fit well within the traditional methods and theorems of mathematics. It is difficult to model complex adaptive systems using Newtonian-style equations.

Computer-based models

Computer-based models handle IF/THEN conditionals with ease. However, in most cases, particular values must be supplied to the variables to execute a computer-based model. That is, each execution is a particular case. In this respect, the computer models are akin to experiments. Many runs may be necessary to gain an overview of the outcomes.

2.4 Three modeling tactics

Whether the model is equation-based or computer-based, there are three distinct approaches to constructing a model. These approaches are distinguished by their objectives.

Data-driven models

The epicycle model and similar models are data-driven models, often called *parametric*. One sets up an array of variables and constants (such as the radii and the rates of rotation of the various circles in the epicycle model) in an attempt to match a large database. The use of data-driven models is the most familiar of the three modeling tactics. The objective is to construct a model that uses data to make predictions. Theories and models for weather prediction are good examples.

Existence-proof models

An existence-proof model is used to show that something is possible. Until the middle of the twentieth century, self-reproduction was held to be impossible for machines. Hence, self-reproduction was used as a distinguishing characteristic of living organisms. Then, in the 1950s, John von Neumann showed that it was possible to construct a self-reproducing machine. (See von Neumann 1966.) The existence of such a machine completely changed philosophical and scientific discussions of "life."

Exploratory models

These models are the least familiar of the three kinds. Typically, exploratory models start with a designated set of mechanisms, such as the various bonds between amino acids, with the objective of finding out what can happen when these mechanisms interact. For example, what protein configurations can be formed from amino-acid bonding mechanisms? Exploratory models have a particularly important role when we first begin to study a complex system, such as a *cas*. Exploratory models often develop into existence-proof models.

What can these different kinds of models offer signal/boundary studies?

The data-driven approach can be useful as a first step in studying signal/boundary interactions, and it provides a routine way of building signal/boundary models. However, such models are tightly tied to particular data sets. It is not easy to extend the models to signal/boundary interactions in other domains, and data-driven models rarely provide insight into general questions about origin of signal/boundary hierarchies or the coevolution of signals and boundaries.

Existence-proof models, by relying on mechanisms, encourage the search for broadly applicable signal/boundary mechanisms. Because existence-proof s/b models emphasize the modes of interaction between mechanisms, they shift attention to "addressing" techniques, such as motifs and tags, that facilitate interaction. Exploratory models show the adequacy or the inadequacy of a set of mechanisms for generating a certain range of signal/boundary observations. Even when the exploratory model generates the observations, that doesn't prove that the mechanisms used are the ones actually present in the signal/boundary system. Typically, new data must be collected to establish the model's hypothesis as correct.

Exploratory models concerned with signal/boundary interactions are the core of this book. In searching for organizing principles, we will examine models that capture essential signal/boundary observations. When such models shade into existence-proof models, they can provide distinctions not previously available, as in the case of the von Neumann model. In particular, an exploratory model can suggest previously unsuspected connections and interactions.

2.5 Detail and rigor

In investigating a cross-disciplinary signal/boundary question, it is crucial to decide what is to be treated as detail in available data. Disparate signal/boundary systems exhibit so many unique features that the only chance of extracting commonalities between systems is to treat many of these features as details. This decision is usually based on the question being posed.

Commonalities in the different data sets pick out the building blocks ("pieces") on which the broadly applicable laws ("rules of the game") depend. Once again Newton provides a clear example. In formulating his simple laws of gravity, Newton set aside the most obvious fact about moving objects: that they come to rest. This property had been taken as axiomatic from the time of Aristotle. Had Newton concentrated on the endless details of how objects come to rest, he would never have answered his question about the existence of a "universal mechanism" (gravity). He relegated "coming to rest" to a special case by stating that a moving body continues in its line of motion at a constant velocity until some force is applied. To account for the everyday observation that objects slow down and come to rest, he invoked the force of friction to distinguish earthbound objects from the planets that move ever onward in their nearly frictionless orbits.

When one is constructing a theory, the inclusion of details can be very tempting. Adding details is especially tempting for computer-executable models, because such models can easily handle large amounts of detail. Indeed, when computer-executable models are presented at conferences, the most frequently asked question is "Why didn't you include parameter x?" These questions stem from the instinct for "realism," which leads to requests for more and more "verifying" details. However, in searching for powerful models, this temptation to inclusiveness should be resisted. A model's clarity and generality depend directly on how much detail has been set aside. The elegance and simplicity of Newton's and Maxwell's equations, though rare, offer clear examples of what can be achieved.

One way to construct an overarching signal/boundary model is to eliminate features not held in common by different signal/boundary systems. By eliminating these details, the model deals with recurrent phenomena—common features that occur and reoccur in the dynamics of different signal/boundary systems. Recurrent phenomena supply building blocks for the model. As a concrete example, consider the flight panel of an airplane.

Each instrument measures some feature that is essential to the control of the plane—airspeed, fuel consumption, altitude, attitude, and so on. These are the recurrent features in the plane's dynamics. They determine its trajectory. If we look to the parts of the plane that generate these readings, we arrive at the plane's building blocks.

Of course, what is detail for one theory may prove to be critical for another. As was emphasized earlier, detail depends on the question being posed. Different questions highlight different mechanisms. Phrasing questions is an art form, not a deductive process. It depends on metaphor, meaning, and other broadly defined human capabilities that, taken together, go by the name "taste."

A brief aside about "rigor," sometimes mistakenly (in my opinion) taken as the opposite of "taste": Although mathematical theories and computer-executable models differ in their procedures, both are completely rigorous. Consequences are formally derived, without the intervention of rhetoric or interpretation. In a sense, computer-based models are *more* rigorous than mathematical models. Every instruction in the program that defines the computer-based model must be correct, and no instruction can be left out. Otherwise the model will not perform as intended. On the other hand, in order to keep traditional mathematical proofs comprehensible, many "obvious" steps in the deduction are left out. In Whitehead and Russell's *Principia Mathematica* (1910), even a simple rigorous proof in arithmetic, such as the proof that addition is commutative, takes several pages when all the steps are filled in. As a consequence, mathematicians give outlines of proofs. This, of course, can lead to errors and false proofs, a not uncommon occurrence in attempts to prove difficult conjectures.

Because both equation-based models and computer-based models are rigorous, I will use the term "mathematical model," or just "model," where the distinction between equation-based theory and a computer-executable model isn't important. Though rigor sets aside rhetoric and interpretation in the *presentation* of a model, the *development* of a model relies heavily on taste, which is acquired through experience and intuition. Scientific papers rarely discuss the developmental stages in model building, though Maxwell did so in his collected papers (1890). As a result, we see only the completed model, not the process used to arrive at the model. In this exploration of signal/boundary models, I will try to present some of the intuitions that lead to the models we examine.

2.6 Existing theory and models

The requirements for a signal/boundary theory are difficult to meet. To see just what the difficulties are, it is helpful to look at some current theories and models that meet one or more of the requirements for signal/boundary theory. (This is a brief scan. In later chapters I will explain the parts of these studies that apply directly to s/b systems.)

(i) The most venerable theory is based on the *Lotka-Volterra equations*, which describe population changes resulting from predator-prey interactions (Christiansen and Feldman 1986). This approach uses differential equations and so taps into a well-established part of mathematics. These equations have three serious shortcomings for present purposes. First, the equations usually assume the organisms are "fully mixed" (random contact based on concentrations), thus forestalling the study of the spatial effects introduced by boundaries. Second, ordinary differential equations do not handle the pervasive conditional (nonlinear) interactions that characterize signal/boundary interactions. Third, the equations make no provision at all for building blocks and the proliferation of building-block combinations.

(ii) The *cellular automata* introduced by Ulam and von Neumann (1966) do provide for spatial distribution. A cellular automaton provides a kind of simple physics with a discrete geometry (like a checkerboard) and a uniform set of laws that hold at each point (square). Think of each point as containing a particle, with the laws telling how that particle is affected by particles in the immediate neighborhood. The interest centers on how particles change as they flow over that network. Cellular automata are adequate for examining many phenomena of interest here, such as self-reproduction and the origin of signals. However, there is little in the way of standard mathematics that applies directly to cellular automata. This lack is especially telling when it comes to studying the formation of boundary hierarchies.

(iii) *Agent-based models* have been a major tool for studying complex adaptive systems in the last 20 years. The models used have included artificial stock markets, models of flocks and swarms, and models of social systems (Epstein and Axtell 1996; Arthur 1997; Mitchell 2009). These models have offered insights into complex adaptive systems, suggesting why markets have bubbles and crashes, why birds flock and fish school, and how cooperation originates. Current work offers progressively deeper insights into distributed control, where there is no central controller or executive (Han, Li, and Guo 2006). The limitations, again, include a limited amount of relevant mathematics and, so far, little provision for agent conglomerates that provide building blocks and behavior at a higher level of organization.

(iv) *Artificial chemistry* models, especially those developed by Walter Fontana (2006), merge the methods of logic with the "billiard-ball" models of elementary chemistry. This merger makes possible sophisticated proofs about the kinds of products that arise from an initial set of reactants. To bring this approach to bear on questions about the formation of signal/boundary interactions, ways must be found to go beyond this chemistry's homogeneity requirement, providing for spatial arrangements and successive, generated enclosures of subregions.

(v) *Neural network* models date back to work by Kleene (1956) and Rochester et al. (1955), but recent applications start with the simulations of Rumelhart and McClelland. Simulated neural networks are constructed of elements, "artificial neurons," interconnected via multiple inputs and outputs. An element produces a signal on its outputs if a weighted sum of its input signals exceeds a fixed threshold. Learning is accomplished by changing weights on the inputs, using some measure of the usefulness of the neuron's output signal. The forte of artificial neural networks is categorization, as exemplified by an early study of past-tense formation in verbs (McClelland and Rumelhart 1986). Neural networks nicely capture the conditional interactions of signals and boundaries, and learning is an integral part of their performance. Moreover, it can be shown (an existence proof) that neural networks with sufficient *interior* feedback loops are computationally complete—they can accomplish any programmable task. Unfortunately for present purposes, most studies of artificial neural networks deal only with *feedforward* networks (which lack loops), so they are not computationally complete. This difficulty is compounded by the absence of guidelines, theoretical or practical, for programming neural networks with many interior loops. There is another intrinsic limitation. It is difficult to study building blocks and the formation of boundaries in this format. Groups of neurons with defined functions can be located in some cases; however, it is difficult to describe mechanisms for recombining the building blocks to form new structures, though Hebb (1949) made a start at this.

(vi) *Classifier systems* (Lanzi 2000) and *Echo models* (Mitchell 1996) were designed to use genetic algorithms to produce adaptation and evolution. Classifier systems have often been used to define the behaviors of agents in agent-based models. Echo models have been used to study evolving ecosystems. Classifier systems allow many simultaneously active IF (signal)/THEN (action) rules, using signals to determine the interaction and sequencing of rules, so they capture the "parallel activity" in signal/boundary systems. We'll see that tags can play an important part in the evolution of classifier systems. So far, however, there have been few studies of the role of tags in defining evolving boundaries in classifier systems, and there is almost no mathematics relevant to tags. Language provides a good test of the relevance of classifier systems to signal/boundary concerns. This use of classifier systems will be examined more carefully in chapter 10.

(vii) *Mathematical network theory* is good at removing details to get at the overall organization of a system. The usefulness of network theory is nicely exemplified by the use of foodwebs to study ecological systems. Most modern network theory is interested in organizational (topological) properties of the network (Newman, Barabasi, and Watts 2006). This interest extends to determining local properties, such as clusters (or communities) of nodes. Though there have been some studies of how this topological organization affects flows of signals and resources over the network, there have been few studies of how evolution and coevolution influence network structure. For signal/boundary studies, it is important to understand how clusters of nodes, for example, become building blocks for the regularly changing topology of the network.

As we go along, we will draw on parts of many of these models. Tags and related ideas from classifier systems will play a particular role in integrating these parts to form a framework for studying signal/boundary systems.

2.7 Requirements for signal/boundary theory

In constructing a general signal/boundary theory, building blocks have high priority because they offer a way to approach the formation of complex signals and boundaries from simple elements. It is encouraging that building blocks underpin the emergence of complexity in all closely examined complex adaptive systems, but we are still in difficult territory. Building blocks alone do not a theory make. Nevertheless, the building-block approach to complexity provides us with a set of four specific requirements for a constructive signal/boundary theory.

Requirement 1

Signals and boundaries, and all things employing them, should be defined with the help of a *formal grammar* that specifies allowable combinations of building blocks.

A formal grammar is a distant, more interesting relative of the grammatical rules we used in grade school to avoid illegitimate sentence structure. Formal grammars, including related formal systems such as axiom systems and production systems, all capture the same basic idea. A small set of *generators* (e.g., the axioms of Euclidean geometry) is transformed by a small set of rules (e.g., rules of deduction) into a large array of objects of interest (e.g., theorems about geometry). Mathematical note: Finitely generated groups provide a rigorous example of mathematical structures defined in terms of generators. Generated systems, as a generalization of grammars, will be discussed in detail in chapter 11, where an elementary exposition of finitely generated groups will be used as a starting point. We can obtain an overarching framework for examining coevolutionary changes in signal/boundary systems by looking at a class of generated systems that centers on the step-by-step dynamics of combining generators (an approach that will be discussed in detail in chapters 13 and 14).

Of course, the generators and the generating rules must be chosen so that the objects generated relate to the questions we hope to answer. A grammar may seem obvious once the task is done, but it is rarely easy in the doing. It took the talents of Mendeleev and many others to arrive at the generated objects of classical chemistry, using atoms as generators and valence rules to determine the possible compounds. For chromosomes, the discovery of the building blocks (nucleotides) and the way they combined was sufficiently difficult to merit a Nobel Prize (Alberts 2007). Similarly, we can generate the set of all ordinary proteins from an alphabet of twenty amino acids, but it is truly difficult to find a "dynamic grammar" for the three-dimensional protein configurations. In each of these cases, the allowed combinations are generated by a grammar that specifies a "vocabulary" (the generators) and the ways of forming "sentences" (the generating rules). For signal/boundary questions, the generators must generate boundaries, signals and, particularly, the signal/

boundary conglomerates that serve as agents. We'll start this search in the next chapter.

Requirement 2

Each generator used by the signal/boundary grammar should have a location in an underlying *geometry*, and combinations of generators should be mobile within that geometry.

Local conditions play an essential role in the adaptation and evolution of complex adaptive systems. Spatial inhomogeneities offer agents diverse opportunities for adaptation and coevolution. In biological cells, spatial inhomogeneity makes possible differing concentrations of reactants at different locations, a factor crucial to cell function. In large-scale complex adaptive systems, such as ecosystems, spatial inhomogeneity arises through variations in the external environment, with different localized patches of resources (water, shelter, food, and so on). In mathematical models, spatial variation is often set aside in the interest of tractability, but spatial variation is a sine qua non when it comes to defining boundaries. By moving from point to point in an inhomogeneous environment, an agent encounters differing opportunities that serve as grist for evolutionary mechanisms. Features that affect ease of movement (e.g., physical barriers, ranging from membranes to mountains) give rise to boundaries that delimit local niches.

Requirement 3

The grammar should be capable of generating *programmable agents*—bounded conglomerates that can execute arbitrary signal-processing programs.

This requirement comes from the very definition of a complex adaptive system: the interaction of large numbers of diverse *agents*. In a biological cell, these agents can be thought of as proteins and other reactants that are conglomerates formed from amino acids and other simple compounds. To define these agent-conglomerates, the generators of the grammar must be a suitable alphabet for defining both the conglomerate-agents and their interactions. In particular, then, the grammar must provide for conditional interactions of the form IF (signal) THEN (new signal). In more general terms, the agents must be "programmable," using the generators and rules provided by the grammar. It is equally important that boundaries be defined using the same alphabet and rules used for defining agent behavior. Moreover, the boundaries should be susceptible to the same mechanisms for change as the other objects in the framework. In short, we should define boundaries, and the resulting agents, as signal-processing conglomerates over a fixed set of generators.

Finding a grammar for programs is not difficult. (Recall that computer programs are built using a simple grammar over a small set of instructions.) However, the requirement for programmability becomes considerably more complicated when the same grammar must be used to generate agent-defining boundaries and boundary hierarchies. The grammar must also provide for the simultaneous execution of different "instructions" at different locations, making it possible for active instructions to send signals that activate instructions at other locations. Boundaries can block signals, and they make possible "loops" of conditional interactions, much like subroutines in a program. By providing locations within the loops, the boundaries act like register addresses in a computer program, facilitating feedback and recycling.

Satisfaction of requirement 3 is a giant step toward adaptive considerations. Identical agent-conglomerates can have distinct individual histories because of conditional reactions to different local environments and agents. The agent's boundary encapsulates this history, so that propagation of the boundary propagates the effects of that history. Behaviors that enhance the persistence of a particular agent configuration favor the future influence of that configuration, as well as adaptations based on that configuration. As an agent-conglomerate becomes more complicated, its structure may actually develop as the agent matures, as in the case of multi-celled organisms.

Requirement 4

The signal/boundary grammar must provide for *reproduction by collecting resources*, whereby an agent-conglomerate reproduces by collecting copies of the generators that define its structure.

Agents acquire generators from the strings of elements that pass through their outer boundary. The elements so acquired can be recombined to form building blocks for a copy of the agent. To provide for this process, the grammar discussed in the previous three requirements must be extended to emphasize using acquired elements to create new boundaries and conglomerates. When the new conglomerates combine to make a new version of the agent, reproduction is implemented. An agent-conglomerate that does not produce such a copy eventually dissolves (dies).

Under this copying process, fitness depends on an agent's ability to acquire the resources it needs in order to make a copy of itself. This version of fitness contrasts with the usual definition of fitness in mathematical genetics, where a predetermined "fitness function" is used to assign a value to each agent. It is difficult, if not impossible, to use a predetermined function to account for the usefulness of a complicated exchange of signals and resources between agents. Yet these exchanges lead to the kinds of coevolution that are familiar from ecosystems: predation, symbiosis, parasitism, and the like. Moreover, under resource-based reproduction there are opportunities to produce unforeseen combinations of building blocks in the offspring agents, further complicating the concepts of agent persistence and fitness.

Summary of requirements

Requirement 4, in combination with the other three requirements, characterizes a signal/boundary theory in which signals, boundaries, and rules are represented by strings generated from a small alphabet. Agents are defined by specifying a boundary hierarchy that contains signals and rules at various levels. The agents are situated in an underlying geometry, and agents are formed out of a spatially distributed population. The generating procedure must provide for the generation of agents with any possible programmable behavior, so that an agent's behavior is not arbitrarily limited by the generating procedure. Then the generating procedure must provide for Darwinian selection in which agents that collect resources more rapidly than others contribute more of their characteristics to future generations. In particular, "subroutines" that appear in rapidly reproducing agents should appear in new combinations in the future.

3 Agents and Signal Processing

3.1 Typical agents in a complex adaptive system

A complex adaptive system (*cas*) consists of a multitude of interacting components called *agents*:

System	Agents	Signals
Biological cell	organelles	signaling proteins
Immune system	antibodies	antigen fragments
Ecosystem	species	sounds, sights, smells
Market	buyers/sellers	buy/sell orders
Language	humans	sound sequences
Government	bureaus	memoranda
Internet	computers	messages

The agents are diverse rather than standardized, and both their behavior and their structure change as they interact. This book, as a whole, aims at increasing our understanding of the signal/boundary mechanisms that give rise to this flexible, changing behavior. Three properties of complex adaptive systems bear strongly on this search for mechanisms of change.

There is no universal competitor or global optimum in a *cas*.
As with the rainforest, a *diverse* array of boundaries and interactions gives rise to niches occupied by clusters of agents. In these niches, *recirculation* of resources can give rise to chains of co-operating specialist agents. The resulting multiplier effect, in which a resource is used over and over again before it dissipates, gives an advantage to co-operating specialists over a holistic generalist processing the same resources. Adam Smith's example of the production of straight pins provides a simple illustration—the coordinated sequence of specialists quickly displaced the single craftsman. It is a reasonable conjecture that recirculation supports increasing diversity of signals and boundaries, but it is a conjecture we will want to examine closely. Whatever the reason, chains of specialists occur in everything from biological cells to economic systems, with signals synchronizing the chains that make efficient interactions possible. Because of these ever-increasing possibilities for interaction, improvements are always possible. Just as it makes no sense to try to define an optimal rainforest, so it is with complex adaptive systems in general. For individual agents, the mechanisms of change aim at improvement rather than optimization.

Innovation is a regular feature of *cas*.
In a complex adaptive system, equilibria are rare and temporary. Adaptation by recombination of "building blocks" is a continuing process at all levels, and *coevolution* clearly has a central role in this continuing adaptation. Any time a new kind of agent appears, there are multiple opportunities for new interactions that modify temporary local equilibria. A single new agent can be at once prey, a partner for exchange, and a parasite, and even more complicated interactions are possible. The new inter-

actions make possible further interactions and adaptations. Because the agents incorporate adaptive mechanisms, the systems continue to innovate. In other words, the mechanisms of change are primarily mechanisms of exploration rather than exploitation.

Lack of a universal competitor and innovation also imply that *cas* agents change their behavior on two different time scales.

Fast

The immediate reaction of an agent to the signals from other agents is conditional, taking a condition/action form. For example, an agent in an equities market might have the rule IF (the market is falling, and my broker says buy) THEN (send a "buy" order). At any specific time, different agents will have different condition/action combinations, resulting in the complex interactions that typify a complex adaptive system. If the situation is changed slightly, the agents respond differently. Moreover, because the conditions involve signals from other agents (as in the broker's message above), there is no possibility of obtaining the action of the whole *cas* by "adding up" the action possibilities of isolated agents—complex adaptive systems are quite nonlinear. As a result, most of the powerful theorems of mathematics, which depend heavily on linearity, are only marginally helpful. In this respect, the mathematical study of complex adaptive systems is quite analogous to the mathematical study of digital computation. You *can* use differential equations to define the actions of a complex adaptive system or a digital computer, but there are no theorems about differential equations that capture such important characteristics as switching, memory, and anticipation.

Slow

Agent adaptation in the evolutionary sense requires the appearance of new agents in the *cas*. As we will see in chapter 6, evolutionary adaptation usually proceeds by introducing new agents formed by recombination of "building blocks" present in extant agents.

In a *cas*, anticipations change the course of the system.
This characteristic entered our earlier discussion of the rainforest, but only in a simple form. The ubiquitous slime mold anticipates periods of scarcity through prepared signal/boundary mechanisms for aggregation of the amoeba-like cells, followed by the formation of spores. Even more simply, the common gut bacterium, *Escherichia coli*, swims up a sugar gradient in anticipation of food. Clearly, anticipation *doesn't* require consciousness, even though anticipation is usually discussed in the context of consciousness. On a human scale, markets can change radically on the basis of anticipations, a change that takes place even if the anticipation isn't realized. In general, anticipation appears whenever an agent, conscious or otherwise, acquires an internal model of its environment. In a *cas*, the environment includes other agents, so the internal model usually includes models of other agents. Then, conscious-like actions begin to appear. Our first discussions of internal models will occur in connection with language in chapter 10.

Human language indeed offers a good test of the relevance of these three characteristics to a *general* understanding of complex adaptive systems because language is a *cas* very different from the examples used to arrive at these characteristics, thus offering new comparisons and tests of signal/boundary concepts; because there is a long history of language studies, with data sets comparable in size and extent to those for rainforests and biological cells; and because language clearly involves complicated conscious actions and anticipations.

The three basic characteristics of a complex adaptive system will be interpreted in detail for language in chapter 10, but here is a brief description of their relevance to language.

Lack of a universal competitor

The diversity of language is well known, as are the niche-like barriers that this diversity presents to human interaction. Different languages have different advantages. Though there have been attempts to determine a universal grammar, there is clearly no universal language in practice.

Innovation

It is well established that modern languages have common roots, but new languages and dialects have arisen regularly over the centuries. Thus, innovation has been a regular feature of language from the outset.

Anticipation

Language involves anticipation in a variety of ways. In a hunting-gathering tribe, the combination of a few words can indicate a food source not directly visible. Language lets the tribe "see over the hill," anticipating the consequences of moving in that direction. As languages became subtler, humans arrived at the notion of "rules" or laws that describe an ever-changing world. The earliest Egyptian dynasties had already invented rule-based board games in which the movements of pieces were governed by arbitrary, language-defined rules. Progressive sophistication of rules led to Thales' logic (J. R. Newman 2003), to Euclidean geometry, and to the theories and models of science. Each advance offered new tools telling us "where to look" for new possibilities, a powerful form of anticipation.

These three characteristics—no universal competitor, innovation, and anticipation—pose different barriers to understanding, but they do help in setting requirements that *cas* agents must meet. By meeting these requirements, we ensure that agents, and the explorations based on them, fit within a broad *cas* framework. Agent-based approaches have been substantially facilitated by computer-based simulations, which now have a richness rivaling that of video games. Agents take conditional actions on the basis of signals generated from outside (by means of a computer keyboard), as well as on the basis of signals generated by other internal agents. The advantages of a rule-based, signal-processing approach to signal/boundary systems are examined in the next section.

3.2 Signal processing

It is important for agent-based signal/boundary studies to keep longer-term, coevolutionary changes at center stage. As was pointed out in the preceding section, such changes are implemented by changes in the rules that allow the agent to respond rapidly and conditionally to signals. However, to see how longer-term rule changes arise, it is necessary to understand how the signal processing can evolve.

The signals used by s/b agents occur at levels ranging from that of the molecules that integrate the activities of a biological cell and the photons that initiate light-sensitive reactions to that of human languages. Because of this range, signals act and change on many different time scales. Moreover, a new way of signaling usually has profound effects on agent-based systems—witness the vast changes in human interaction that stem from the Internet. The origin of language, or at least the extensive growth and reorganization of the primate brain that made language possible, may have been triggered when primates encountered a rich new network of signals and boundaries at the savannah/forest edge. Whatever the cause, language offers a clear, ever-expanding signaling method that has greatly altered the world of primates and, indeed, the planet. To compare signal processing in these different contexts requires a uniform way of representing different signal-processing rules.

Classifier systems

Classifier systems (Lanzi 2000), discussed briefly in chapter 2, provide a class of formally defined, rule-based signal-processing systems useful for defining agents. Individual rules in the system, called *classifiers*, are of the form

IF (a required set of signals is present)

THEN (send an outgoing signal based on these signals).

Thus, signals determine both the interaction of classifiers and the order in which classifiers are executed, mimicking the sequencing of instructions in a digital computer. In addition, classifier systems allow many rules to be active simultaneously, allowing signals to be "broadcast" throughout the agent. Classifier systems have the further advantage that they were designed to undergo adaptation and evolution using genetic algorithms, thus pointing ahead to evolutionary concerns (discussed in chapter 5).

For simplicity, the signals in classifier systems are usually represented as binary (bit) strings over the alphabet {1,0}. This restriction to binary strings doesn't cause a loss in general applicability, because any string (say, a string of amino acids) can easily be recoded as a binary string. It is important that the signal sent by a classifier has no intrinsic meaning—it is simply an uninterpreted bit string. A signal's only effect occurs when it satisfies one of the conditions of another rule, contributing to its activation. In this, the signals are much like the pulse sequences inside a computer chip. Because the signals have no intrinsic meaning, they cannot "contradict" one another. The broadcast of multiple signals simply means that more rules can become active. A fortiori, any subset of rules can be active at the same time because their actions, the sending of signals, cannot cause contradictions. This multiplicity of signals is analogous to the presence of many signaling proteins in a cell.

Classifier rules gain their flexibility by using a simple method to classify signals and define conditions. A rule condition is specified by augmenting the signal alphabet {1,0} with an additional symbol #, yielding a condition-defining alphabet {1,0,#}, where a condition is just a string over this augmented alphabet. The condition, so specified, accepts any signal that matches, place for place, the ones and zeroes in the condition, ignoring the positions where # occurs, so the # is a kind of "don't care" symbol. For example, the condition 1##### accepts any signal that starts with a 1; it "doesn't care" what occurs in the rest of the signal. Similarly, the condition #0#11# accepts any signal with a 0 in the second position and ones at positions 4 and 5, such as the signals 100110 and 001111. More formally, a condition C accepts any signal satisfying the following recipe:

At each position at which condition C has a 1, the signal must have a 1 at the corresponding position.

At each position at which condition C has a 0, the signal must have a 0.

At all other positions the signal may have either a 1 or a 0.

A condition is said to be *satisfied* by a signal if it accepts that signal. I will adopt the convention that a condition C ending in a # accepts *any* signal of *any* length if the first part of the signal matches the part of the condition prior to the terminal #. For instance, the condition C = 000# will accept a string of any length that starts with 000, such as 000 or 0001011 or 000111 . . . 11, whereas the condition C = 000 will accept only the string 000 of length 3.

For ease of presentation, I will limit examples to classifiers with two conditions and a single response:

IF (a signal of type a is present) & (a signal of type b is present)

THEN (send an outgoing signal d).

Because a classifier rule with multiple conditions and responses can be easily implemented by setting up a cascade of two-condition classifiers, the exposition that follows holds for rules with more than two conditions and more than one response.

The signals broadcast when many rules are active at the same time are collected on a *signal list.* For example, if the classifier system is modeling the set of proteins in an organelle at a particular time, the strings corresponding to those proteins will be collected on the signal list. Phrased in terms of the signal list, a classifier rule with two conditions has the form

IF (the signal list has a signal of type a) & (the signal list has a signal of type b)

THEN (place signal d on the signal list).

Each classifier in the system, at each time step, checks the signal list for signals matching its conditions; if those signals are present, the classifier adds its own signal to the list (for use on the next time step).

A classifier system, then, consists of a list of rules and a signal list. At each time step, all rules simultaneously check the signal list. Rules that have their conditions satisfied post their signals to the signal list for the next time step. Signals are ephemeral, lasting only one time step. To keep a signal on the list for more than one time step, there must be a rule that continues to post it. Under this arrangement, the set of signals on the signal list at any time is the set produced by active rules on the preceding step (including signals produced by rules responding to agent's

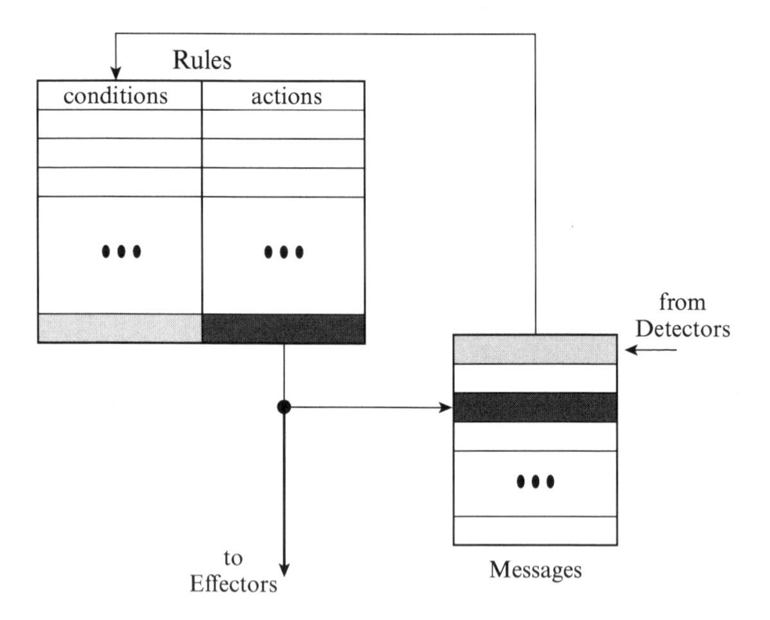

Figure 3.1
A classifier system.

environment). Accordingly, the signal list gives the current state of the classifier system—the rules use only this list to determine the next state of the system.

Classifier systems were designed to allow this "parallelism" of many simultaneously active rules. If we add more rules to the system or delete rules from the system, the behavior of the system will change; however, its behavior will always be clearly defined, without any contradictions that must be resolved. In effect, we have a *population* of interacting rules. That, in turn, means that adaptive mechanisms similar to those in population genetics can be used to cause the classifier system to evolve. Indeed, genetic algorithms, when applied to classifier systems, provide a computer-executable model of populations undergoing natural selection.

Genetic algorithms

Genetic algorithms are based on the classic view of a chromosome as a string of genes, and they work with populations of strings. R. A. Fisher (1930) used this format to found mathematical genetics, taking a generation-by-generation view of evolution in which, at each stage, a population of individuals produces a set of offspring that constitutes the next generation. A *fitness* function assigns to each string the number of offspring it will produce. In this "string of genes" format, each gene on the string is drawn from a fixed set of alternatives, called *alleles*. For classifier systems, the alleles are {1,0,#}.

The genetic algorithm, starting from an initial population, produces successive generations using the following subroutine:

(0) Start with a population of N individual strings of alleles (perhaps generated at random).

(1) Select two individuals at random from the current population, biasing selection toward individuals with higher fitness.

(2) Use genetic operators (such as mutation and crossover, described below) to produce two new individuals, which are assigned to the next generation.

(3) Repeat steps 1 and 2 *N*/2 times to produce a new generation of *N* individuals.

(4) Return to step 1 to produce the next generation.

Natural selection, in this formulation, can be thought of as using genetic operators to search through the set of possible individuals—the *search space*—to find individuals of progressively higher fitness. The most familiar genetic operators are mutation and crossover. Mutation replaces an allele with a randomly selected alternative. Crossover (often called recombination in genetics) is defined simply. Two chromosomes are lined up, then a point along the chromosome is randomly selected, and pieces to the left of that point are exchanged between the chromosomes, producing a pair of offspring chromosomes:

$A_1A_2A_3\ A_4A_5|A_6A_7 \ldots A_k \quad B_1B_2\ B_3B_4B_5A_6\ A_7 \ldots A_k$

=>

$B_1B_2\ B_3\ B_4\ B_5|B_6B_7 \ldots B_k \quad A_1A_2A_3A_4A_5\ B_6\ B_7 \ldots B_k$

To a good first approximation, this is the form of crossover typically observed in mating organisms. Crossover takes place in every mating, whereas mutation of a given gene typically occurs in less than one individual in a million. Clearly the high observed crossover frequency gives it a preeminent role in evolution. This is the central reason that mating individuals produce offspring exhibiting a mix of their parents' characteristics—

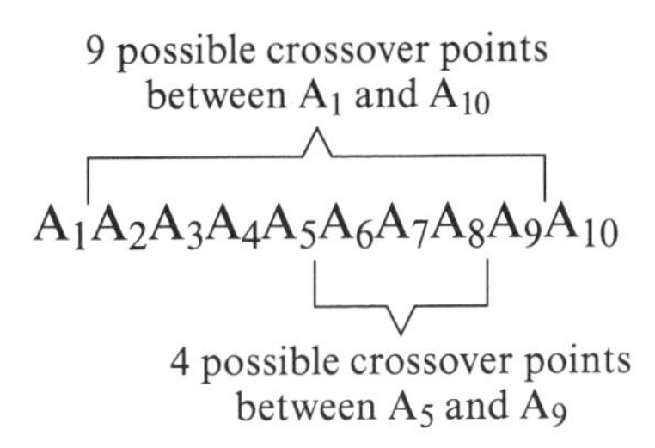

Figure 3.2
Linkage.

consider the human population for a vivid, familiar example. Likewise, crossover is used in artificial cross-breeding of selected animals and plants to produce superior varieties.

It is noteworthy that, with a randomly chosen crossover point, alleles that are close together on the chromosome are likely to be passed as a unit to one of the offspring, whereas alleles that are far apart are likely to be separated by crossover, one allele appearing in one offspring and the other allele appearing in the other offspring. For example, for a chromosome with 1,001 genes, the chance that an adjacent pair of alleles will be separated in the offspring is 1/1,000, whereas alleles at the ends of the string will always be separated. In standard genetic terminology, this phenomenon is called *linkage*.

Despite the simple definition of crossover, the phenomenon of linkage produces considerable complication in the study of successive generations. Because of crossover, a highly effective individual in the parent generation (say, an "Einstein") will pass only a subset of its alleles to any given offspring. This raises an important question: If a parent's particular arrangement of alleles is never passed on, what is preserved from generation to generation? One answer to this question turns on a prediction of the generation-by-generation spread of *clusters* of alleles. Such a prediction requires a substantial generalization of Fisher's fundamental theorem about the spread of individual alleles. That generalization will be discussed in chapter 6.

When the genetic algorithm is used with classifier systems, the population consists of strings representing the rules of the system. Then the genetic algorithm, generation by generation, searches for improvement within the set of possible rules. When rules of above-average fitness share well-linked clusters of alleles, crossover rapidly uncovers and exploits those possibilities.

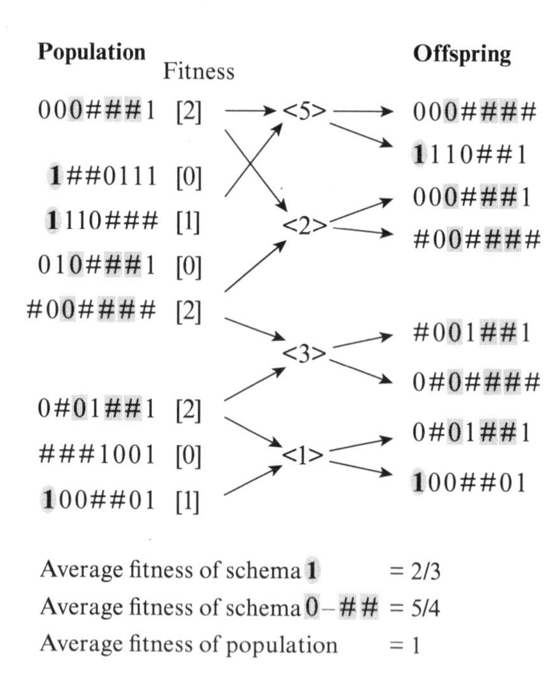

Figure 3.3
Effect of the genetic algorithm on a population.

Situated agents

An agent is *situated* when it is placed in an environment that consists of other agents and "inanimate" objects such as obstacles and patches of resources. To interact with this environment, the agent must have a way of detecting local features of the environment and a way of modifying that environment. For an animal-like agent, detection would depend on the familiar senses (sight, sound, touch, and so on); action would depend on various combinations of muscles, producing everything from movement to speech.

For *cas* signal-processing agents, we will invoke a set of *detectors* for determining the changing features of the surrounding environment. To preserve the binary form of signals in a classifier system, each detector responds to the presence (1) or absence (0) of some particular property. For example, one property might be whether or not there is a "moving object" in the nearby environment. An array of eight detectors for a simple situated agent could then be as follows, with the detector on the first line and the observed property on the second:

d_1	d_2	d_3	d_4	d_5	d_6	d_7	d_8
moving	*airborne*	*black*	*large*	*winged*	*fuzzy*	*multi-legged*	*long*

The detector array serves as an interface that transforms properties of the environment into a binary signal string that serves as input to the situated agent.

Each *effector* of a situated agent defined by a classifier system is activated by an appropriate signal on the agent's signal list. Many effectors can be active simultaneously when the appropriate signals are on the signal list. Though the signals themselves cannot conflict, the actions of different effectors *can* conflict. For example, one effector could cause the agent to "turn right" and another could cause the agent to "turn left." If simultaneous signals try to activate both effectors, the ensuing conflict must be resolved. There are several ways to do this. The simplest, and least interesting, is to choose one of the conflicting actions randomly. A more interesting approach, and one that will be examined later, uses a notion of signal strength—the effector receiving the stronger signal is the one activated. Because a typical agent has many fewer effectors than rules, it is much easier to resolve conflicts between effectors than it would be to resolve contradictions between rules. By leaving signals uninterpreted, we avoid the complex problem of resolving internal contradictions (such as occur in an "expert system" in which rules are hypotheses about "facts").

3.3 Reactions

Individual rules in a classifier system, like the individual instructions for a computer program, are generally simple and easy to understand. However, a system with many interacting rules is as difficult to analyze as a computer program with many subroutines. After decades of working with computer programs, we still have few powerful tools for analyzing them. There is, however, a path that helps in analyzing classifier systems. By relating classifier rules to the reactions of elementary chemistry, we can bring into play some well-studied models from elementary physics and chemistry.

All the interactions in a biological cell—certainly a complicated signal/boundary system—satisfy the laws of chemistry. Thus, our first step is to consider the classical two-component reactions, called *binary reactions,* that we were exposed to in elementary chemistry:

a + b => c + d.

If we think of the compounds as signals, this can be rewritten as a classifier rule:

IF (a) & (b) THEN (c) & (d).

It is sufficient to consider reactions involving two reactants, because reactions involving more than two reactants can be represented by "piling up" a sufficient number of binary reactions. In other words, if we think of chemical reactions as the result of "collisions" between molecules, then simultaneous three-way collisions are rare, and any such collision can be thought of as the result of a pair of two-way collisions.

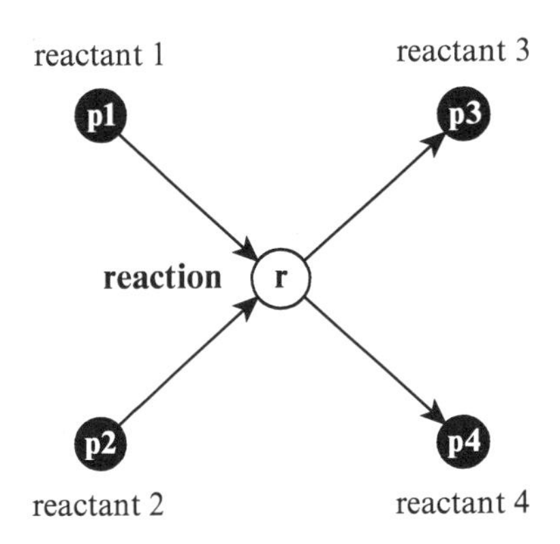

(p) reactant with concentration **p**

(r) reaction with rate **r**

Figure 3.4
A binary reaction.

Recalling that proteins are defined by strings of amino acids, we can define proteins a and b as the strings

$a = a_1a_2 \ldots a_r$

and

$b = b_1b_2 \ldots b_s,$

where each letter in the string is an amino acid. In typical protein-protein interactions the component amino acids are not modified by the reaction; the amino acids are just recombined to produce two new strings—say,

$c = a_1a_2 \ldots a_jb_k \ldots b_s$

and

$d = b_1b_2 \ldots b_{k-1}a_{j+1} \ldots a_r.$

Often the recombination is more complicated than this simple exchange, involving several subsets, but the total number of copies of each amino acid is still conserved. In short, we are considering reactions in which there is a "conservation-of-letters" requirement. This conservation requirement contrasts with the usual implementation of a signal-processing system, in which the signal's letters are simply "plucked out of thin air," as is the case for classifier systems. However, this additional conservation constraint is satisfied for most signal/boundary interactions. Moreover, as with classifier systems, an appropriate set of reactions can carry out any computation if sufficient elements are supplied to the input. Thus, we retain the power, and the generality, of being able to describe any set of signal/boundary interactions with this more constrained binary reaction system.

Interestingly, most reactions between proteins are determined by relatively small subsequences, called *active sites*, within the string defining the protein. Different proteins with the same active sites typically enter the same reaction(s). From the string-processing point of view, the subsequences defining the active sites play of the role of *tags* (mentioned in section 2.6). That is, the subsequences play an "addressing" role, directing the proteins into certain reactions. The actual three-dimensional form of the active site depends on the other, "structural" parts of the protein string, but its "addressing" function is then fully determined. As was mentioned earlier, it may be difficult in practice to determine the active site's functional form because it is difficult to go from the protein's amino-acid sequence to its three-dimensional structure (a problem known as the *folding problem*; see Alberts 2007). Fortunately, that level of detail isn't necessary for the questions we are examining here.

Looking at the tag-like properties of active sites provides some useful insights. Consider a reaction in which two proteins,

$a = a_1a_2 \dots a_r$

and

$b = b_1b_2 \dots b_s$,

are "spliced" to produce a single product protein:

$c = a_1a_2 \dots a_rb_1 \dots b_s$.

Let the two proteins have active sites $x = a_1a_2a_3$ and $y = b_1b_2$ respectively. Then the condition part of the reaction takes the form

IF (x#) & (y#),

where, by convention, a condition x# with a terminal # accepts any string with header (tag) x. Other proteins with the headers x and y—say,

$a' = a_1a_2a_3a'_4a'_5 \ldots a'_{r'}$

and

$b' = b_1b_2b'_3bb'_4 \ldots b'_{s'}$

—will also enter this same splicing reaction, but the result is a different product:

$c' = a_1a_2a_3a'_4a'_5 \ldots a'_rb_1b_2b'_3bb'_4 \ldots b'_{s'}$

Note that c' need not even be the same length as c.

This dependence of the product on the reaction's input can be handled in the classifier-system format by a rule of the form

IF (x#) & (y#) THEN (x#y#),

where x#y# on the THEN side of the rule indicate that the letters following the tags x and y in the input are *passed through* to the output in the same order in which they occurred in the input. Specifically, if the first reactant is $a = xa_4a_5 \ldots a_r$, then $a_4a_5 \ldots a_r$ is passed through; if the first reactant is $a' = ya'_4a'_5 \ldots a'_r$, then $a'_4a'_5 \ldots a'_r$ is passed through. Similarly for b and b'.

This splicing rule can be thought of as a kind of catalyst that acts whenever it encounters reactants with tags x and y. In biological cells, catalysts often accelerate reactions by a factor of a thousand or more, so that, on the time scale of the cell, the effect is that of a go/no-go for the reaction. In other words, if the catalyst (rule) is present the reaction takes place, otherwise it takes place too slowly to have an effect on the cell's metabolism. The next section will look at this catalytic effect of rules in a simple chemistry, a chemistry based on collisions of reactant particles.

The reverse of a splicing operation is a splitting operation in which a is split into two parts, c and d. To define the reaction we must specify the point at which $a = a_1a_2a_3a_4a_5 \ldots a_r$ is to be split. We could think of having a semicolon at the split point:

$a = a_1a_2a_3 \ldots a_h ; a_{h+1}a_{h+2} \ldots a_r$.

The split then results in

$c = a_1a_2a_3 \ldots a_h$

and

$d = a_{h+1}a_{h+2} \ldots a_r$.

The semicolon can be implemented as a split-specific tag (say, $x = x_1x_2x_3$), so that x plays the role of a semicolon in

$a = a_1a_2a_3 \ldots a_h\ x\ a_{h+1}a_{h+2} \ldots a_r$.

The split operation then results in

$c = a_1a_2a_3 \ldots a_hx$

and

$d = a_{h+1}a_{h+2} \ldots a_r$.

The corresponding rule-based form is

IF $(a_1a_2a_3 \ldots a_hx\#)$ THEN $(a_1a_2a_3 \ldots a_hx)$ & (#).

Similar considerations hold for the general binary reaction

$a + b <=> c + d$.

As an example, consider an *exchange* reaction in which an initial segment of a is spliced to a terminal segment of b and vice versa. As in the preceding paragraph, let

$a = a_1a_2a_3 \ldots a_hxa_{h+1}a_{h+2} \ldots a_r$.

and

$b = b_1b_2b_3 \ldots b_jyb_{j+1}b_{j+2} \ldots b_s$.

The results of the exchange are

$c = a_1a_2a_3 \ldots a_hxb_{j+1} \ldots b_s$

and

$d = b_1b_2b_3 \ldots b_jya_{h+1} \ldots a_r.$

The rule-based form is

IF (ax#) & (by#) THEN (by#) & (ax#).

Note again that the segments designated by the terminal # are passed through in the order given in the conditions so that the first segment $a_{h+1}a_{h+2} \ldots a_r$ is attached to by# and vice versa.

candidate reactants

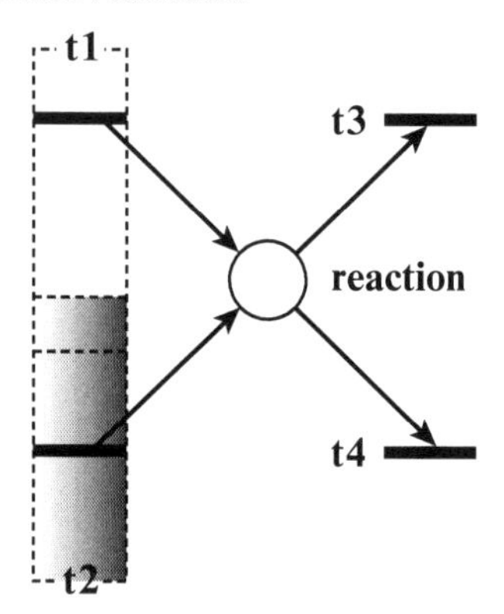

t1, t2, t3, and t4 are *tags* (ligands, receptors, active sites).

▬ is a particuar string; a string **t1**a1...ak is a condidate for first reactant; a string **t2**b1...bk is a candidate for second reactant.

This binary reaction can be represented as a classifier rule:
t1#...# & **t2**#...# ⟶ **t3**a1...ak & **t4**b1...bk

Figure 3.5
The role of reaction tags.

At the cost of adding a new *pass through* symbol, (||), this rule could be made more obvious:

IF (ax#) & (by||) THEN (ax||) & (by#).

Tags can occur anywhere in the reaction string, in analogy to the fact that active sites in a protein can occur at several locations along the string of amino acids. Indeed, from the point of view of a classifier rule, a tag can be thought of as any location in the rule's condition that doesn't have a # or a ||. Thus, a condition in the form ###00# would treat the values 00 at the fourth and fifth positions as a tag. A condition in the form ###00###111# can be interpreted as looking for two tags: 00 and 111.

This relation between classifier rules and reactions offers an approach to a basic question that has a pivotal role in signal/boundary investigations: What coevolutionary mechanisms give rise to a signal/boundary system consisting of many reaction rules, boundaries, and signals? The witches' brew mentioned in chapter 1 allows a great many reactions. What kinds of greasy globs and tarry gunk can we expect? If we change the mixture slightly, what will happen? Similar complexities attend other signal/boundary systems. The next section takes a first step toward answering these questions. By using tags and related ideas from classifier systems, we move from the particularities of elementary chemistry to the general realm of agents with complex signal/boundary hierarchies.

3.4 Particle (billiard-ball) mechanics and the concept of state

The conservation-of-letters requirement for reactions lets us exploit a well-studied model from classical physics: *particle mechanics* or, less formally, billiard-ball mechanics. In billiard-ball mechanics, each reactant is treated as a particle or a billiard ball, different reactants being identified by different "colors." In a two-dimensional version, the billiard balls move at random on an idealized, frictionless billiard table. Under this regime, the balls collide at random, as happens with the molecules of a hot gas. When two balls collide, a reaction may take place. The reaction is determined by the colors (reactant types) of the colliding pair. Two outgoing balls, the products of the reaction, replace the incoming balls, each product being assigned the appropriate color. In effect, the colors of the two colliding balls are changed by the reaction.

As a simple model of this kind of chemistry, consider a bowl filled with Mexican jumping beans that are continually jumping (Asimov 1988). Some jumps are high and some jumps are low, with higher jumps less likely. For present purposes, let the jumps become exponentially less likely as the height increases, so that a jump of height 2h is $1/(2^2) = 1/4$ less likely than a jump of height h. (This exponential decrease mimics the exponential decrease of thermal energies in a gas, as specified by the tail of the famous Maxwell-Boltzmann distribution; see Feynman 2005.) In a simple particle chemistry, the height of the jump corresponds to the energy available to the particle for a chemical reaction. In the bowl analogy, the chemical reaction takes place if the particle escapes from the bowl. Roughly, a catalyst corresponds to putting a notch in the bowl (a point that will be considered more carefully in the next chapter). If the bowl's height is 2h and the notch is at height h, then balls that manage to hit the notch will escape with four times the frequency of balls that do not hit the notch.

To a good approximation, proteins enclosed within a single membrane encounter one another uniformly in the fashion of a billiard-ball mechanics. From a rule-based point of view, uniform distribution means that all rule (reaction) outputs are broadcast uniformly to all other rules within the boundary. In other words, uniform distribution corresponds to a classifier-system signal list making all signals available to all rules.

Under uniform mixing, encounters are determined by the relative proportions of the reactants in the local mixture. If there are more balls of color b then more collisions will involve b; if *all* the balls are of color b then *all* collisions will involve b. For the reaction a + b => c + d, the rate of production of c and d, then, depends on (i) the probability that a will encounter b and (ii) the proportion of collisions between a and b that result in the products c and d (the probability that a and b will react corresponds to the reaction rate, which might be zero).

More carefully, if the overall proportions (concentrations) of a and b are p_a and p_b, the probability that a will encounter b is given by $p_a p_b$. Probability ii, the *forward reaction rate*, is given by a reaction-specific constant $r_{ab|cd}$. Multiplying these two probabilities gives the probability that a collision of a with b will yield the results c and d:

$$p_c = p_d = r_{ab|cd}\, p_a p_b.$$

In elementary chemistry this is the familiar formula for the rate of production of the products of the reaction. We will use it to advantage when we begin examining the effect of boundaries in signal/boundary systems in chapters 7 and 8. For now, simply note that a boundary, by selective admission, can greatly change concentrations within its domain, thereby favoring certain reactions, products, and interactions.

System states

To help capture the complex possibilities inherent in different mixtures of reactants, let us turn to a concept that is fundamental to the study of systems: the *state* of the system. Board games provide a helpful analogy for understanding this concept. The state of a game at any time is the placement of game pieces on the board. The set of all possible placements that can result from play of the game is the game's *state set* S. Similarly, the state set S of a system is the set of all configurations of the system's elements (particles, reactants, or the like) that can occur under the "laws" of the system (laws of motion, reaction, or the like). Just as the play of the game amounts to moving from board configuration to board configuration according to the rules of the game, so the *dynamics* of a system amounts to moving from state to state according to the laws of the system.

In most games, the play from any board configuration B onward *doesn't* depend on the moves leading to configuration B. What is "legal" from configuration B onward depends only on B, not on how you got to B. The same is true in physics: knowing the state of a system is sufficient for determining future possibilities. Consider "billiard-ball mechanics" again. The state of the reaction system is given by the concentrations of the reactants. Knowing these concentrations is enough to determine the frequencies of different reactions, and that, in turn, determines new concentrations of the reactants. As with the board configuration, we can determine the new concentrations from the current concentrations, without knowing how the current concentrations came to be.

The concept of state provides a uniform way to describe the dynamics of diverse signal/boundary systems, and it will aid us in our search for an overarching framework. In particular, we can use the state concept (chapter 7) to relate binary reaction systems to a part of elementary probability theory that has been used to model everything from games of chance to Mendel's experiments with peas. In chapter 8 this model will be extended to membranes and non-uniform mixtures. But, first, it is helpful to have a uniform way to represent the *interactions* of an arbitrary conglomeration of agents, be they antibodies or business firms. Networks, formally defined, offer this possibility; they are the subject of chapter 4.

3.5 Signal processing in a biological cell

In a biological cell, most signals are proteins, and are structurally determined by strings over the twenty-letter amino-acid alphabet. To a good first approximation, these proteins are uniformly distributed *within the compartment or structure containing them*, and they interact in random collisions (though the rate of their collisions may be slowed down because the compartment is crowded). Encounters between proteins, and the resulting reactions, are controlled in two ways: via selective admission of proteins to a compartment and via catalysts called *enzymes*. By selective admission, compartment boundaries can cause internal reactant proportions to differ greatly from proportions in the environment surrounding the boundary. When that happens, the compartment favors reactions of the highly concentrated interior proteins. Enzymes, proteins specified by the cell's genes, augment this control by catalyzing selected reactions. As was mentioned in section 3.3, on the time scale typical of cell metabolism a reaction that isn't enzyme-mediated is unlikely to have a role in the metabolic network. Enzymes, then, act as rules for putting together and splitting proteins; that is, they are signal processors.

The biological cell points up a feature that will be a central in the signal/boundary framework presented here: that all protein-specified signals are defined by strings over an alphabet of twenty amino acids. It is significant that enzymes (the "rules") are also proteins, defined on the same alphabet. Moreover, subsets of those strings—active sites, motifs, and the like—act as tags determining the interactions of the proteins carrying the tags. In short, the agents in the cell can be specified as strings over a small, fixed alphabet. Succeeding chapters will show that this is also true of the other signal/boundary systems.

3.6 Summary

Bringing the mechanisms of billiard-ball chemistry into the classifier-system signal-processing format provides a way of using well-studied mechanisms to understand the full array of signal/boundary interactions. Because classifier systems are computationally complete, any computer-executable model can be captured within the resulting framework. Moreover, classifier systems are specifically designed for use with genetic algorithms, so they provide mechanisms for the adaptive coevolution of signals and boundaries. Before we consider adaptation (in chapter 5), a closer look at the networks generated by a population of reactions will enable us to look beyond details to pervasive signal/boundary structures that serve as grist for the adaptive mill.

4 Networks and Flows

4.1 Networks

Part of the difficulty in understanding the dynamics of complex adaptive systems is the mass of detail that hides common patterns in the flow of signals and resources. Network theory, more formally called *graph theory*, offers a general way of describing arbitrary *cas* interactions by extracting the pattern from the details. Serious study of networks, which began in the 1930s (Konig 1936), has undergone a recent renaissance (Newman, Barabasi, and Watts 2006). Networks were briefly discussed in chapter 1 under the topics *niche and hierarchy* and *recirculation*; now let us look at them more carefully.

Foodwebs are simple networks that capture the "who eats whom" interactions in ecosystems (Levin 1999). They are easily understood, exhibiting both the possibilities and the shortcomings of the network approach. In a foodweb, there is a node for each organism being studied, with a directed connection from that node (organism) to each organism it consumes. (See figure 4.1.) The data needed to construct such a network are usually readily available, at least for larger fauna and flora. To keep the

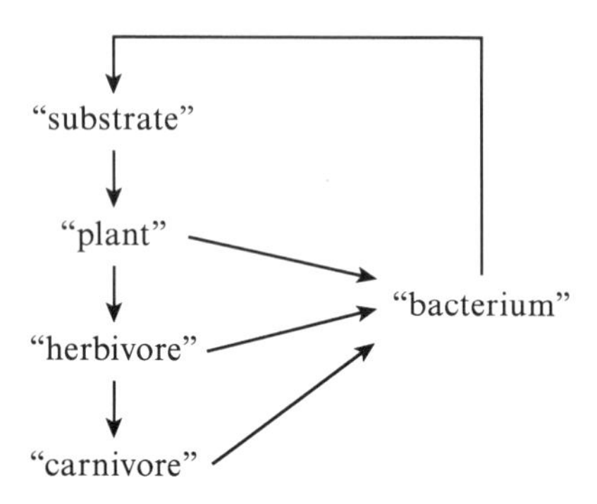

Figure 4.1
A foodweb.

network simple, foodweb designers usually set aside variations in the availability of prey over different portions of the landscape, so that connections are fixed. These simplifying assumptions strongly restrict applications, but they make mathematical analysis and modeling possible.

To see one way a foodweb model can be used, consider this question: How easily can a new organism invade an established ecosystem? To investigate this question, we start with a foodweb describing the established ecosystem. Then we modify the foodweb by adding a new node that represents the invading organism. The question is then rephrased: Will sufficient resources flow through the new node to enable the invading organism to persist? When the behaviors of the organisms being examined fit the simplifying assumptions for the "who eats whom" network, this rephrased question can be answered mathematically. Furthermore, generality is gained. The answers can be applied to other systems in which consumption has a central role, such as a network of buyers and sellers. In more complex cases, where mathematical approaches are difficult, foodwebs can still serve as a basis for computer-executable models.

Note that change has a role even in this seemingly simple question about invasion. The addition of the node representing the invader changes the network and the flows over it. In the *formation of niches* (boundaries) within the foodweb (considered in chapter 9), change moves to the center of the stage. Indeed, extant niches in realistic systems are rarely fixed for long—they are continually forming and re-forming. Even in simple cases, the prey eaten by a predator depends on the prey's availability in the vicinity of the predator. If the favorite prey is in short supply, there can be a sudden increase in the consumption of other prey, with corresponding changes in the flows over the network. For instance, when orcas ("killer whales") find their favored prey, seals, in short supply, they suddenly put heavy pressure on populations of sea otters. The lion cannot eat the lamb if the lamb isn't present. Both orcas and lions are flexible enough in diet, and mobile enough, to shift to other prey. Unlike a mature tree, they can scour a large territory. Such biogeographic effects, including territorial influence, migration, over-predation, and seasonal change, obviously force us to go beyond static networks to an examination of mechanisms that change networks.

Changes in the interactions within a niche can be subtle. Consider the following example of *phenotypic plasticity*. Frog tadpoles usually forage at the top of a small pond, consuming large amounts of algae. Dragonfly larvae are fierce predators on tadpoles. Over time, evolution has provided tadpoles with a "wired-in" anticipation. When they detect the scent (pheromone) of dragonfly larvae, they switch to foraging in the dead leaves at the bottom of the pond, where they are less obvious targets. In more technical terms, in response to changing conditions in the niche, they change the phenotypic activities that define their niche behavior. When the tadpoles forage at the bottom of the pond, the algae at the top of the pond flourish, because the freely moving dragonfly larvae are carnivores and aren't interested in algae. It isn't easy to represent this interaction with a standard foodweb. And there is a further complication—a kind of cosmic justice: mature frogs are predators on mature dragonflies. Again, we see dramatic changes over time that aren't easily captured by a static network.

A survey of ecosystem data shows that spatial and temporal shifts in prey consumption are a major features in the formation and dissolution of biological niches. It is difficult to ask sharp questions about shifting niches when predators and prey are treated as uniformly available in space and time. Accordingly, we must concern ourselves with dynamically changing, *spatially distributed* networks. Unfortunately, networks with continually changing connections have been little studied, so this poses a problem for using networks as a ready-made vehicle for signal/boundary theory. For this reason, networks will serve us more as a way of sharpening intuition than as a basis for theory. Despite the limitations, networks can be interpreted over the full range of signal/boundary interactions, enabling comparison of ideas and tools from quite disparate fields. The objective now is to develop the relation between networks and the signal-processing agents of the preceding chapter.

4.2 Tags

Tags, the address-like regions in signals (see section 3.3), can be used to set up the connections in signal/boundary systems. In analogy to the way active sites determine a protein's reactions, tags determine which rules in a system process the tagged signal. Reciprocally, the condition for a classifier rule can treat any region in a signal as a tag. For example, a condition of the form ###00##111# can be interpreted as looking for strings having both the tag 00 and the tag 111. Accordingly, changes in tags determine changes in the signal/boundary network, giving us a direct way to study changes in these networks.

In particular, tags and conditions based on them supply raw material for adaptive mechanisms that manipulate addressing, motifs, and other building blocks of networks and signal flows. As a simple example, consider cross-breeding a condition C of the form ###00###### with a condition C′ of the form #######111# (used, say, by the exchange reaction defined in section 3.3). Then cross-breeding can yield one offspring that requires both tags and another offspring that requires neither tag:

```
###00#|#####          ###00##111# (more specific)

      |        =>

######|#111#          ########### (more general).
```

Thus, by operating on tags in signals and conditions used by established signal/boundary interactions, cross-breeding and other adaptive mechanisms can produce new, redirected signals and conditions. Such redirection, by using interactions that already have proved useful, can often usefully augment established signal/boundary interactions.

Exploring the possibilities of adaptive mechanisms working on tags (discussed in more detail in chapters 5 and 6) can considerably increase our understanding of signal/boundary coevolution. It is particularly important that tags can impose constraints similar to those imposed by boundaries. Consider the simple case of three rules—C = 11011#, C' = 110011#, and C" = 1101#—with a common prefix tag: 11. If the prefix tag 11 is uniquely assigned to signals originating from a cluster of rules elsewhere in the system, it becomes a kind of name for the originating cluster. Each of the three rules then makes an additional requirement on the signals originating from that cluster, using the follow-on tags ##011#, ##0011#, and ##01#. These follow-on tags act as additional selective requirements on signals coming from the cluster, in effect designating subclusters.

Tags can also be used to force sequential execution of rules, as in a computer program. Consider a set of tags x1, x2, x3, . . . , and a sequence of rules r1, r2, r3, . . . , each with a single condition and a single outgoing signal. Let the outgoing signal of rule rj have tag xj, and let the condition of rule j + 1 require a signal with tag xj. If rj is the only source of signal with tag xj, then rule j + 1 can become active only when the rule rj is active on the previous time step. In this way, we force a sequence of rule executions. We can even implement rule sequences that loop back on themselves, and, by using rules with multiple conditions, we can set conditions under which the loops become active. Implementing subroutines using conglomerates of rules is an important step toward satisfying the requirement that an overarching signal/boundary framework be able to implement arbitrary agent strategies.

In sum, then, tags serve as building blocks for parsing the coevolution of signal/boundary interactions. A tag lets us trace a sequence of changes through a common pattern. In genomics, common DNA sequences (motifs) let us uncover evolutionary paths connecting different species such as determining the earliest common ancestor of a group of organisms. Similar methods have proved useful in studies of the evolution of language, enabling us to determine the earliest form of a group of related languages. Tags lift this approach to signal/boundary systems in general, serving as "markers" for coevolutionary paths in signal-processing systems.

4.3 Representing networks with classifier systems

Tags demonstrate their importance, and their flexibility, when we use a unique binary number to name each node in a network of interacting rules. The directed edges that connect the nodes are then named by a number pair specifying the origin and the destination of the edge. Figure 4.2 (on page 92) depicts a network of six nodes and six directed edges. The nodes are labeled 0000, 0001, 0010, 0011, 0100; the directed edges are labeled by pairs indicating origin and destination: (0000,0001), (0001,0010), (0001,0011), (0010,0000), (0100,0011), (0100,0101). Six classifier rules can represent this network, using one rule for each edge. If node y receives a directed edge from node x, the rule corresponding to this edge has the form x# / y#. The five-node graph of the figure then has the following representation:

Edge	Rule
(0000,0001)	0000# / 0001#
(0001,0010)	0001# / 0010#
(0001,0011)	0001# / 0011#
(0010,0000)	0010# / 0000#
(0100,0011)	0100# / 0011#
(0100,0101)	0100# / 0101#

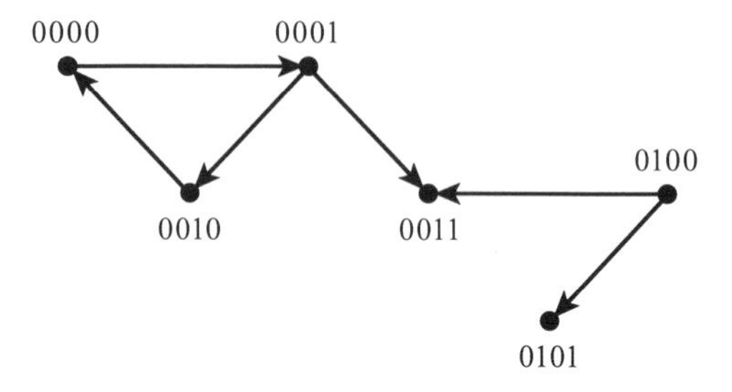

Figure 4.2
A six-node labeled graph.

By using unique tags for each node in the network, we obtain a unique set of rules. Networks represented by rules of this kind automatically process *groups* of signals because of the effect of the terminal #. (Recall that a single terminal # in a condition means that the condition will accept any signal tagged with the prefix, whereas a terminal # in the action part of the rule simply passes through to the corresponding part of the incoming signal.)

Note that all the nodes in the example above have the same initial bit: 0. Thus, 0# can be thought of as a tag that identifies the nodes in this particular subnetwork; a second distinct network could use the tag 1#. Using prefixing tags in this way, we can number whole networks much as we did the nodes. For example, 010011 could be used to indicate node 3 (011) in network 2 (010). It then becomes possible to add rules that connect distinct networks. Indeed, we can even represent subnetworks by assigning the nodes in the subnetwork a common prefix. This, in turn, lets us represent hierarchies within the network. All these possibilities add potential for adaptive modification of the networks.

4.4 Representing reaction networks with classifier systems

The use of classifier-system rules to define reactions in a billiard-ball chemistry was discussed in the preceding chapter. The go/no-go nature of biological catalysts (enzymes) emphasized in section 3.4 gives a natural interpretation for rules assigned to nodes in a reaction network. Each rule plays the role of a catalyst; if the rule is present, the corresponding reaction takes place, otherwise it doesn't. From a billiard-ball point of view, when the reactants collide with a ball that represents the catalyst, the reactants are transformed to products.

Placing classifier systems and tags within a network formalism brings us closer to evolution-sensitive ways of defining "membranes." Following the example of section 4.2, a set C of classifiers using the same tag y in each member's conditions creates the counterpart of a semi-permeable membrane that admits only signals carrying tag y. Similarly, the tags carried by the outgoing signals of the rules in C determine how the reaction products disperse.

The next step is to mimic the membrane hierarchies that control the reactions in a biological cell. The object is to have a straightforward way of describing the effect of "membrane hierarchies" inside a "pot" of reactants, where each "membrane" bounds a community of reactants, hindering their free diffusion. That constraint suggests employing the technique described at the end of the last section, using tags to designate distinct networks. Consider, then, interactions generated by of a set of rules R that produce signals carrying a tag y (see figure 4.3), and another set of rules R' where each rule in R' has a condition requiring signals with tag y. If tag y is used only by R and R', then only signals processed by R can be further

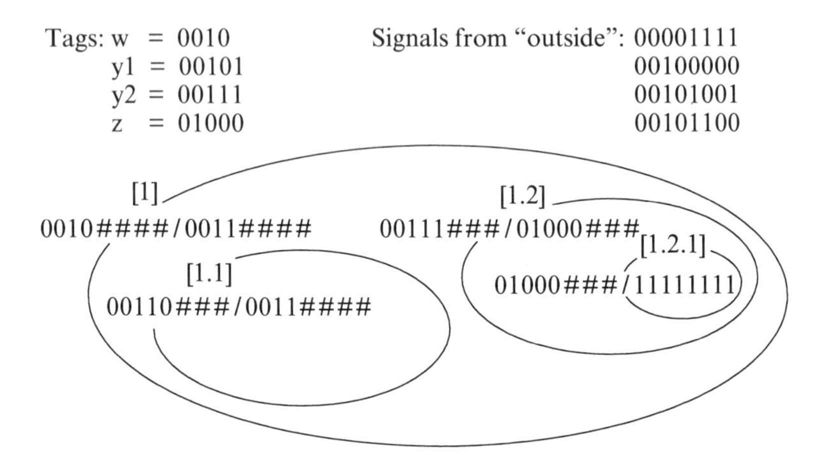

Rule [1] processes the three outside signals with tag w to produce signals 00110000, 00111001, 00111100. If the tags are unique to these rules, then [1.1] can only process signals first processed by [1]; that is, [1.1] is *interior* to [1].

Figure 4.3
A tag-generated boundary hierarchy

processed by the rules in R'. That is, R' is "interior" to R with respect to signal processing.

Obviously, we could then assign a unique tag y' to the outputs of rules in R', and add a set of rules R" that require tag y' for input. Then R" would be interior to R', which is already interior to R. So begins a hierarchy of boundaries that can be extended indefinitely through a sequence of tags y, y', y", ... Because tags, signals, and rules are defined over a single alphabet in classifier systems, it isn't difficult to find mechanisms that generate tag sequences in classifier systems. That is, in classifier systems there is a direct way to generate boundary hierarchies. Moreover, because we can define reactions using classifiers, we can extend this approach to reactions in general (chapter 7). We can then go on to model the origin and adaptation of boundary hierarchies within a signal/boundary framework.

At this point it is worthwhile to look again at the four general properties of signal/boundary systems discussed in section 1.3: *diversity*, *recirculation*, *niche and hierarchy*, and *coevolution*. Because of the conservation-of-elements provision, elements (letters) are passed down rule sequences. This pass-through makes multiplier effects possible, favoring specialization and, ultimately, a *diverse* array of specialists. When the sequence bends back into a loop, we see *recirculation*. The use of tags to provide for boundary hierarchies provides the necessary means for defining *niche and hierarchy*. That leaves the *coevolution* property still to be handled.

An examination of evolution and coevolution in networks goes well beyond typical network studies. It requires examination of adaptive mechanisms for changing network structure, a very different operation from the conditional actions associated with the nodes. Most agent-based studies involve the interactions of agents situated on a fixed network using fixed IF/THEN rules. The relatively few network studies of adaptive agents still use a predetermined network ("evolutionary graph theory"; see Nowak 2006). However, the coevolution of agents requires a different approach. We must look at networks (e.g., foodwebs) that change as the agents co-adapt. Changes in the exterior boundaries of the agent change its possibilities for interaction with other agents; changes in its interior boundaries change its interior signal processing.

The changes in networks generated by co-adaptation pose questions that involve aggregate or global network properties: Is there increasing recirculation as the network evolves? Does the network become less (or more) vulnerable to changes occurring in other parts of the network? Does the processing associated with the nodes become increasingly diverse? The usual tools of network theory are not well suited to answering such questions, because simple rules for generating the changes—say, connecting new nodes to nodes already highly connected (Newman, Barabasi, and Watts 2006)—aren't explicitly geared to Darwinian mechanisms for adaptation. The changes generated by a Darwinian process must directly affect the persistence of a node's connections over time.

In this chapter, in preparation for exploring the changes induced by coevolution, we have moved from the formalism of elementary chemical reactions to the general realm of agents with complex signal/boundary hierarchies, using tags and related ideas from classifier systems to approach general networks of boundary-mediated interactions. Using tags, we can examine the coevolution of agents by examining the coevolution of the tags. (In the next chapter, tags will be used as grist for adaptive mechanisms that drive agent coevolution.)

4.5 Networks in biological cells

Metabolic networks are used to describe the flow and transformation of proteins within a biological cell (Alberts et al. 2007). The atoms from which the proteins are constructed (primarily carbon, nitrogen, hydrogen, and oxygen) are conserved as the proteins are transformed within the network. This conservation of elements imposes a kind of budget that can be used to account for the flow of the atoms as they pass from reaction to reaction. Much as with cash in a sophisticated economy, these elements pass through diverse activities with substantial recycling (as in the Krebs cycle). Multiplier effects arise (recall Adam

Smith's example), favoring specific, efficient reactions. Highly specific enzymes are one result of this specialization.

The cell's counterparts of tags—motifs, active sites, etc.—determine the routing of proteins through the metabolic network. Though we have increasingly detailed specifications of metabolic networks, those specifications rarely lend themselves to studies of the coevolutionary steps that led to the network's origin. The routing function of tags opens the way to studying these steps by studying coevolutionary modifications of the tags while still exploiting the advantages of networks in describing interactions. In biological cells, genetic mechanisms modify parts of genes that specify tag-like sections of proteins, providing rerouting within the metabolic network.

5 Adaptation

5.1 Adaptive changes in interactions

Chapter 4 made the point that in order to understand signal/boundary systems we must consider the *formation* of boundaries and signals, not just their existence. In complex adaptive systems in general, and in signal/boundary systems in particular, adaptive mechanisms mediate the formation of new structures, often adding specialized versions of existing components to an agent. As a result, flows of resources and signals in the agent pass through increasing numbers of stages, each of which encompasses some special activity of the original flow.

More than 200 years have elapsed since Adam Smith highlighted the flow of resources in pin production, but we are still far from a good understanding of the mechanisms that generate this transformation. What is difficult in this special case must, a fortiori, be difficult in the broader realm of signal/boundary transformations. Nevertheless, Smith's pin factory illustrates three major characteristics of signal/boundary interactions: that new boundaries distinguish the specialists, that resources flow from specialist to specialist, and that signals synchronize interactions.

In the formation of new boundaries, signals influence the structure of new boundaries at least as often as boundaries influence the formation of new signals. Changes in both signals and boundaries occur on many different time scales. Changes in biological cells are typically rapid (in human terms), but there are many different rates, ranging from nanoseconds for chemical reactions to hours or more for cell division. Political boundaries and their associated governments change over years, the transformation of Serendip to Sri Lanka being of literary note. Geographical changes sometimes occur within a human life span (the formation of oxbow lakes is an example), but most take millennia. As Lyell's *uniformitarian principle* emphasizes, geological changes occur over the still longer time spans typical of the evolution of species (Allen 1977). Are there adaptive mechanisms that operate over all these time scales?

5.2 Mechanisms of change (as suggested by biological cells)

In line with the discussion of progressive enclosure in section 4.4, we must look for mechanisms that generate networks with two properties: (i) a hierarchical arrangement of nodes, reflecting the hierarchy of boundary enclosures, and (ii) connections distinguished ("colored") by the signals they carry.

Here again I'll turn to biological cells for insight, but I reemphasize that the interest here centers on mechanisms that extend beyond biology to signal/boundary interactions in general. The preceding chapter concentrated on three of the four signal/boundary features of highlighted by tropical rainforests: diversity, recirculation, and niche and hierarchy. The object here is to take a closer look at the fourth of those features: coevolution.

Within the outer membrane of a biological cell, there are various distinguished structures, such as the DNA complex found within the nucleus (Alberts et al. 2007). These distinguished structures mediate varieties of signaling, feedback, and catalysis seen only in the most complex signal/boundary systems. The resulting hierarchy of boundaries and signals determines everything from what genes are active to what is transported through the membranes. To sort out the mechanisms that underpin this continually changing network of interactions, let's start with the five W's of the news reporter: Who, What, Where, When, and Why.

Who

The "who" in this case are the chromosomes and proteins that act as *agents* within the cell. Both chromosomes and proteins are constructed of sequences of "letters" from small alphabets—four nucleotides (A, G, C, and T) for chromosomes, and twenty amino acids for proteins. For both chromosomes and proteins, the primary properties are determined by the order of the letters—different sequences yield different reactions. The nucleotide sequence for a chromosome is typically hundreds of millions of letters long. For example, the human genome is made up of more than 3 billion pairs of nucleotides (Alberts et al. 2007). The chromosome is, famously, a double helix, and this simple structure becomes the backbone for a complex involving many additional attachments. Despite the attachments, the classical notion of the chromosome as a sequence of genes, with each gene being a subsequence of the overall nucleotide sequence, supplies a useful overview. Humans have on the order of 20,000 genes that encode proteins.

A closer look at chromosomes reveals several levels of organization. Classically, there were two levels: the nucleotides and the bounded sequences of nucleotides that define genes. However, there is an interesting intermediate level that codes for parts of proteins called *motifs*, amino-acid sequences that appear in many different proteins (Rhind 2011). Protein motifs are similar to musical motifs, serving as building blocks that allow proteins to combine into recognizable structures that go by names such as "alpha helices" and "beta sheets." There is also a level of organization above the gene that consists of gene "subroutines" coding for coordinated protein cascades and cycles. One of the best known of these "gene subroutines" codes for the cycle of protein catalysts called the Krebs (or citric acid) cycle, which codes for the precursors of the amino acids that are sequenced to form proteins. This tiered organization, in which building blocks at one level combine to yield building blocks at another, is a theme we will examine closely, beginning in the next section.

To add to the program-like nature of chromosomes, genes can be repressed ("turned off") and induced ("turned on"). When a gene is on, it is "decoded" in such a way that it sets off a cascade of actions that result in the construction of specific proteins; genes that are off are not expressed. Thus, the same chromosome can have different effects in different cells, depending on which genes are expressed. The result mimics the execution of a sophisticated computer program with many interior loops.

Proteins, in contrast to DNA's double helix, fold into complex three-dimensional configurations that determine the properties of the proteins. Within a biological cell, proteins determine the permeability of membranes, act as signals, and function as catalysts. Proteins acting as catalytic *enzymes* can increase specific reaction rates by factors as large as 10,000 (Alberts et al. 2007). On the time scale of a cell, the presence of an enzyme can make the difference between a reaction that affects the cell's function and a reaction that has little or no effect. Again we encounter a conditional procedure: IF (catalyst present) / THEN (reaction takes place). When a protein acts as a signal that turns a gene on or off, it provides feedback from parts of the reaction network that generated that protein. Genes newly turned on then provide new proteins, which can turn other genes on or off. Gene subroutines so activated can constrain the concentrations of selected proteins (negative feedback) or cause precipitous changes in concentrations of others (positive feedback).

There is a long mathematical tradition, *population genetics*, that studies the evolution and spread of genes (Christiansen and Feldman 1986). However, nearly all the models used in population genetics treat the effects of active genes as additive, ignoring the feedback provided by gene subroutines. As was pointed out earlier, the problem of understanding the interactions of chromosomes and proteins is similar in form and in difficulty to the problem of understanding the interaction of instructions in a computer program. Under these circumstances, treating gene effects as additive is of little avail. We have to look beyond the methods of population genetics to model gene subroutines and their evolution.

What

What form(s) do these complex interactions take? We know that all the reactions in a biological cell satisfy the laws of chemistry.

To a first approximation, all the compounds at a given location—say, within an interior membrane—are uniformly mixed. From the discussion of reactions in section 3.4, we know that we can then use the classic formula

$$p_c = p_d = r_{ab|cd}\, p_a p_b$$

to determine the probability that products c and d result from a collision of compounds a and b. The discussion of reaction models in chapter 4 emphasized the importance of tags as grist for adaptation. In chapter 7, this elementary reaction model will be expanded into a model that handles membranes and nonuniform mixtures.

Where

In a biological cell the localization of reactants provided by interior membranes is critical. The cell uses the successive enclosures provided by semi-permeable membranes to construct its own local variations in concentrations of reactants. (Here and hereafter I will use "semi-permeable" to designate boundaries that constrain the movement of molecules, either through "pore size" or by actively "pumping" selected molecules through the membrane.) Because of the many interior membranes (boundaries), and because of spatial separation, the reactant concentrations in the cell vary strongly from location to location, despite uniform mixing within the interior compartments. The localization provided by the interior membranes, in turn, permits local specialization of reactions and signals. In effect, boundaries separate the system into individual agents, which retain distinct individual histories (experiences).

Once we have different histories for different enclosed sets of reactions, differential adaptation via natural selection becomes possible. Enclosures that persist over longer periods are more likely to take part in further adaptations, so there is selection for reaction sets that increase their own persistence. Increased persistence can be favored directly (say, through enzymes that act on reactions in the same cycle that produces them, called *autocatalysis*) or indirectly (as with recycling in cascades or "production lines" involving other enclosed sets). Clearly niches and specialization have a role to play here, a theme that will be examined carefully in chapter 9.

When

There is a continually changing kaleidoscope of interactions in a biological cell—biological cells regularly divide to produce offspring, and there is a broad pattern of successive interactions leading from the birth of an offspring cell to the time the new offspring cell undergoes division itself. This succession is controlled by means of the chromosomal "program" that determines the subroutines active at each of the stages leading to division. Many coordinated interactions take place simultaneously during these stages. The cell is never in a "stable" state, so the network that describes its interior chemical reactions is continually changing.

Why

In view of the great diversity of reactants in a biological cell, the number of possible reactions is very, very large. At first sight, it seems that the result should be a kind of "witches' brew" with gunk of all kinds. But that isn't the case. The IF/THEN nature of actual reactions, with catalytic proteins increasing reaction rates by large factors, yields tight control. Each effective reaction is completely dependent on a specific set of precursor reactants and enzymes. Thus, it is necessary to discuss the precursor conditions in order to discuss the question "Why did that interaction take place?"

The question then becomes "Why are those precursor reactants present, and why is that enzyme present (or absent)?" The precursor conditions are primarily determined by the selective actions of membranes and by signaling proteins from other parts of the cell. The membrane-induced distortions in reaction concentrations and the catalytic actions induced by the signals both strongly favor certain reactions over others. The result is a chain leading backward from precursors to precursors of the precursors and so on. In this process there will be many feedback loops, but ultimately the chains must trace back to inputs to the cell and to the genes being expressed, or repressed, in the chromosomes.

Once again, tags have a major role in defining active sites, membrane pores, and protein signal destinations. Adaptive mechanisms use tags as building blocks to be combined and recombined, the tags persisting when they enter chains that increase the cell's fitness. Thus, the coevolution of tags provides an adaptive process that is much more directed, and more plausible, than random variation.

Summary

This quick review of the "who, what, where, when, and why" of the cell shows us that great adaptability can come from relatively simple underpinnings. Though the interactions in biological cells are complex, the structures governing the interactions are simple. Sequences over two simple alphabets are "decoded" to construct the proteins that define most of the structures in the cell: active sites, membrane pores, and signaling. Moreover, a limited set of bonding rules for chemical elements determines the decoding operations. Despite these simplicities, it is quite difficult to determine what activities have been encoded. The relationship between the code and the behavior is comparable to the relationship between a sequence of computer instructions and the computation it implements.

Researchers who decode protein functions find recurrent patterns in proteins (motifs, tags) that are associated with particular kinds of activities. The role of these recurrent patterns is analogous to the named patterns that recur in chess—patterns with names such as "fork attack," "pin," and "gambit." In chess, these patterns form building blocks for strategies that suggest useful ways to continue the game. It is instructive that we still discover new, useful patterns in chess even though the game has been around for centuries. So it is in the adaptive process of discovering and exploiting tags.

Because a cell contains large numbers of proteins that are simultaneously active, innovations can often be tested without greatly disturbing processes already in place. Heritable innovations in a cell usually occur through simple operations that modify the chromosomes. Although mutation (random change of a nucleotide) is often cited, recombination—and exchange of segments between a pair of chromosomes—is much more frequent and more effective. Recombination offers a way of generating potentially useful innovations from tested building blocks. Cells employ a simple mechanism called *crossover* for generating recombination. Crossover appears in biological contexts ranging from the immune system to germ cells, and it is used in computer programs (e.g., genetic algorithms; see Mitchell 1996). It will be examined carefully in the next chapter.

These observations of change and adaptation in biological cells are helpful in formulating specific requirements for a signal/boundary theory, and in the next chapter they will be used for this purpose. But first it is useful to look at building blocks in a wider context.

5.3 Building blocks

The emergence of complexity from simple laws or rules is more common than it might seem. The game of chess is defined by fewer than a dozen rules, yet no finished tournament game duplicates any previous game, and, I have already noted, new techniques for playing chess are being found after hundreds of years of study. Much the same can be said for the five axioms of Euclidean geometry; after two millennia of study, geometers are still discovering new theorems. More relevant to our current concerns, the "machine code" of a contemporary computer chip usually involves 32 or 64 basic instructions, and a program is simply a sequence of these instructions. Alan Turing (1936) combined mathematical insight with mathematical theory to give us a principled way of finding *computationally complete* sets of instructions—sets of instructions that, subject to constraints of memory size, can be sequenced to define any conceivable algorithm. In a similar way, the vast and bewildering array of chemical reactions observed by alchemists became organized and, in principle, predictable once we had Mendeleev's periodic table of the elements and their "valences." (See J. R. Newman 2003.) In each example, the system is synthesized by combining a simple, fixed set of building blocks: rules, axioms, instructions, or elements.

Building blocks play a role in analysis as well as in synthesis. The role of building blocks in analyzing visual images was mentioned briefly in section 1.2. The precise light pattern on the millions of cells in the human retina is never twice the same in a human lifetime. However, there are recurring patterns that allow us to see commonalities in a bewildering array of nonrepeating sensory patterns. We extract coherence by thinking of the world in terms of combinations of building blocks (features, objects, and the like), which, unlike the overall retinal light patterns, do recur. On an entirely different time scale, by analyzing weathering effects on persistent geological patterns, such as mountains and rivers, Charles Lyell founded the modern analytic science of geology (Allen 1977). He adopted a *uniformitarian* hypothesis. Effects observed over a few years, such as weathering and erosion, would continue at the same rate over longer periods. By looking at mountains in various stages of erosion, and river deltas in various stages of extension, it was possible to infer the time involved. It quickly became apparent that erosion and delta extension had been going on for tens of millions of years or more, and that Earth was much older than had previously been thought.

Both innovation and diversity result from combining familiar building blocks in new ways. The internal-combustion engine was constructed from components that had been familiar for many decades. A similar comment holds for the transistor. (See Goldberg 2002.) When building blocks have already been tested in other contexts, new combinations often turn out to be workable or viable, as in the case of biological cells. Innovation through recombination of building blocks leads to sustainable diversity.

In short, by extracting building blocks and examining their combinations we acquire a way to understand a wide spectrum of complex adaptive systems, ranging from biological cells and visual perception to games, computer programs, and inventions. In most complex adaptive systems, building blocks at one level of complexity are combined to get building blocks for structures at a higher level of complexity. This layered use of building blocks can be compared to the combination of nucleons to get atoms, which in turn are combined to get molecules, and so on. This plausible, experience-based nature of layered structures formed of tested building blocks again holds whether we are talking of evolution, innovation, or mathematical proofs.

5.4 Two kinds of building blocks

Building blocks are often not at all evident when we first look at some complex adaptive systems. This is so even though we have observed thousands of different structures that, later, we find can be constructed from a set of simple building blocks. Whole ranges of chemical reactions had been observed and used for thousands of years (consider the chemistries of glass, clay, and metal) before Mendeleev compiled the periodic table. To use building blocks to increase our understanding, we must confront the question "How do we extract building blocks from unfamiliar systems?"

Clearly, whatever we choose for a building block must be bounded in a way that lets us distinguish it from its surroundings. This distinguishing boundary typically acts as a filter or barrier that limits the ways a building block can interact with other blocks. Thus, a search for barriers or filters in a *cas* often guides the search for building blocks. Moreover, in a *cas* the building blocks of interest are processors—they carry out conditional actions, as with protein reactants or computer instructions. It is conditional action, based on the context provided by other building blocks, that produces complex behavior from simple building blocks, a point that the next chapter will discuss more carefully. But first there is an important distinction to be made: a distinction between generators and conglomerates.

A building block is a *generator* if it is immutable over the span of its existence. It may be copied and it may be combined with other generators, but it has a fixed set of "connectors" subject to a fixed set of connection rules, as in the case of chemical elements. If a generator is removed from a structure, the structure is changed, typically to a structure with a very different function or interpretation. Removal of a generator is often disastrous, as in the removal of a single instruction from a computer program.

A building block is a *conglomerate* if it can grow and "fission" into additional conglomerates related to the "parent" conglomerate. It is possible for a conglomerate to split into two components, one of which retains the processing capability of the original conglomerate and one of which assumes a new function. In most cases, a conglomerate can be defined in terms of an interconnected set of generators. Then, growth consists in the addition of new, immutable generators to the cluster, and fission consists in separating the generators into two interacting subsets. The pattern that defines the conglomerate usually doesn't persist as long as its component generators. However, a conglomerate isn't of much interest unless it persists long enough to serve as a building block for more complex structures. The Krebs cycle is an example of a conglomerate that acts as a critical building block at a higher level—this conglomerate of eight enzymes appears in the metabolic networks of all aerobic organisms (Alberts et al. 2007). Reaction loops, involving recycling and feedback, can increase the conglomerate's persistence. In addition, copies of the conglomerate can evolve and adapt over longer intervals, much as a biological species can change and adapt. The growth and the fission of conglomerates will be discussed at several points in later chapters; the formation of default hierarchies, discussed in the next chapter, provides an early example.

This distinction between generators and conglomerates is often ignored by researchers, but it can be important. Currently, in brain scans, it often assumed that an observed active region is directly responsible for an associated behavior. That is, observed active regions are treated as generators rather than as conglomerates. On this interpretation, there should be a loss of function when the active region is deleted or seriously damaged. However, it known experimentally that other active regions often "take over" when such a loss occurs. This strongly suggests that the region observed be treated as part of a conglomerate. Similarly, though reaction sequences in a biological cell are often treated as either present or absent, adjustments and takeovers do occur.

Because a conglomerate has parts, it can become a source of new conglomerates through recombination with other extant conglomerates. When the new conglomerates act as building blocks for new adaptations, the process can be repeated, producing still further adaptations, resulting in a system that is perpetually changing. In such circumstances, the conglomerate is best interpreted as a persistent pattern imposed on a flow of building blocks. The human body, usually treated as a static array of organs, is in fact a pattern imposed on a flow of atoms—

its constituent atoms are all "turned over" in less than two years, and most of the atoms are turned over in a matter of days or weeks. Conglomerates, as patterns imposed on flows, can be modified to augment properties such as self-repair and homeostasis (Ashby 1952), thereby further increasing their persistence. Disturbing such conglomerates is much like disturbing the standing wave that builds up before a rock in a whitewater river—as soon as the disturbance is removed, the standing wave reforms. That is, the pattern (conglomerate) is self-repairing.

To repeat an important point: Conglomerates can usually be described in terms of changing combinations of building blocks drawn from the next lower level of description—neurons in the case of the brain, reactions in the case of the cell, and so on. Accordingly, conglomerates will be discussed in terms of their constituent building blocks wherever that is possible.

5.5 A word about emergence

Emergence is a topic closely associated with the study of complex systems—some will not call a system complex unless it exhibits some form of emergence. But care is required in using this term. Emergence is most easy to understand in the context of the physical sciences. For example, consider the rules for the combination of atoms, as reflected in the periodic table. These rules constrain the possibilities for molecules, but the molecules actually observed are only a fraction of the possible combinations. There are rules at the next level—the rules of chemistry—that constrain the reactions and combinations observed. The rules of chemistry use additional information—temperature, pressure, concentrations, etc.—provided by the context in which the reactions occur. The rules of chemistry, however, cannot violate the rules imposed at the atomic level. In other words, the rules at lower levels constrain the rules at the level being investigated.

Emergence in this format comes from patterns or properties that appear under the constraints imposed by the rules of combination. In complex adaptive systems, emergent properties often occur when coevolving signals and boundaries generate new levels of organization. Newer signals and boundaries can then emerge from combinations of building blocks at this new level of organization. Indeed, some properties can emerge only when the system reaches a high enough level of organization. It makes no sense to talk of the wetness of an atom, but atoms combined into collections of molecules can exhibit the aggregate phenomenon we call wetness. Similarly, the production of sound becomes spoken language only at a high level of neural and muscular organization.

The view of emergence as a phenomenon generated by combining building blocks contrasts with the view of emergence as a holistic phenomenon. A holistic phenomenon, by definition, cannot be reduced to an interaction of parts. Clearly, the studies here allow for reduction. However, the reduction goes beyond the traditional reduction wherein "the whole is *equal* to the *sum* of its parts." The conditional interactions between signals and boundaries cannot be simply added up. When the conditional interactions are included, reduction serves as a powerful tool for understanding emergent properties of signal/boundary systems.

5.6 Questions

This general discussion of adaptation and emergence puts emphasis again on the bounded entities called agents. The organelles of biological cells act as agents capable of active processing. With this agent-oriented view, we can recast questions about adaptation as questions about building blocks for agents.

How do agents arise?
What are the building blocks for the boundaries that distinguish the agents?

How do agents specialize?
How does the agent acquire the boundary characteristics that allow it to react to the signals and resources produced by other agents?

How do agents aggregate into hierarchical organizations?
What makes it possible for one agent to include another?

Emphasis on building blocks transforms the foregoing questions into questions about combining and recombining building blocks. The next chapter examines mechanisms for combination and recombination.

6 Recombination and Reproduction

The object at this point is to examine how mechanisms for combining building blocks (generators) can generate a wide spectrum of coevolving signals and boundaries. In particular, Darwinian selection, acting on a signal/boundary system, can ensure that only building blocks residing in fit structures survive to be tested in other structures. When a building block is tested and survives in a variety of contexts, its usefulness becomes well confirmed in a statistical sense. The procedures examined here, rather than treating individual building blocks independently of one another, place strong emphasis on context provided by other building blocks. Ignoring context would be tantamount to ignoring the conditional interactions that highlight signal/boundary interactions.

6.1 Recombination of rules

For thousands of years humans have cross-bred animals to produce offspring that combine favorable characteristics, some of which are present in one parent and some in the other. For example, mating horses that have long legs with horses that have strong necks will yield some offspring with both long legs and strong necks. This recombination occurs naturally in all mating organisms; for example, each of us has some features of one parent and some features of the other.

Recombination occurs because the chromosomes that control development pair up, and one chromosome within a pair physically crosses over the other, exchanging segments on one side of the *crossover point*. The crossover point, to a first approximation, occurs at a random point along the chromosome. A simple example shows the effect of this crossover in producing a recombination of building blocks. A human face can be described in terms of building blocks by picking ten features that are common to all faces: chin, lips, nose, eyes, and so on. Allow ten variants for each of the ten features. (See figure 6.1.) There are 10 × 10 building blocks in all, but by choosing one block for each feature we have a distinct description for any of 10^{10} different faces. Note also that the use of building blocks lets us describe a complex geometric object by a string of ten digits, each digit designating a particular choice for the corresponding building

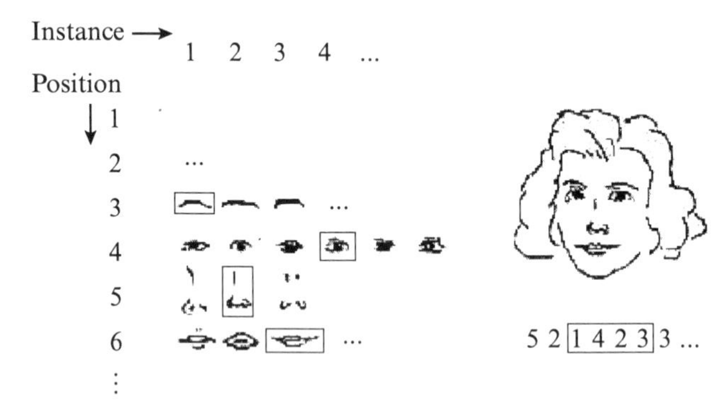

Figure 6.1
Building blocks for a face.

Using the number string representation for faces, new faces can be constructed by using the *crossover* operator on pairs of faces.

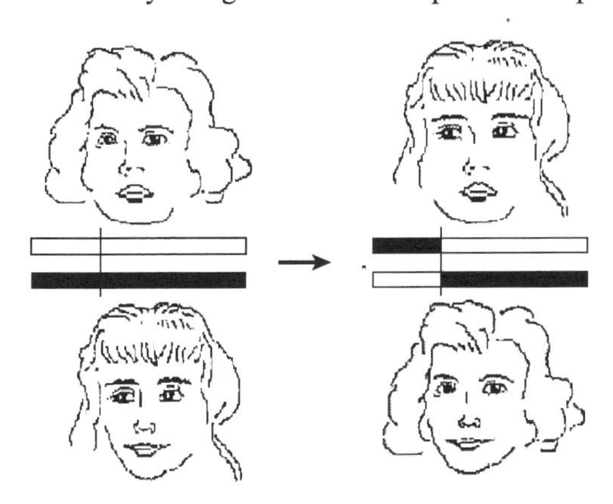

Figure 6.2
Crossover of face building blocks.

block. Two of these strings can be crossed, just as with chromosomes, to yield two new objects.

The random location of the crossover point has some interesting consequences. The chance of the crossover point falling between two alleles that are close together is small—geneticists say they are closely linked. On the other hand, alleles that are far apart are likely to be separated in the offspring. In other words, alleles that are closely linked are likely to be passed in a cluster to offspring, whereas alleles that are widely separated are likely to appear in separate offspring. This is well illustrated in the example of the faces, where clusters of features are passed on to the faces produced by crossover. Long before we had modern micro-techniques, this linkage effect allowed geneticists to use breeding experiments to map the ordering of genes on a chromosome (Lindsley and Grell 1967).

All 10^{10} faces resulting from choices from the 100 building blocks are realistic. This illustrates the possibilities for recombination when the building blocks are well chosen. When the initial population contains a good sample of relevant building blocks, then under recombination subsequent generations will test a broad range of plausible possibilities. With a population of tested classifier rules, crossover can produce plausible new rules in a similar way. The new rules can be thought of as hypotheses to be tested. They are confirmed or disconfirmed via Darwinian selection—good hypotheses survive into later generations.

In examining the effect of crossover on classifier rules, recall first that the more #'s there are in a classifier condition the wider the range of signals it accepts. For example, the condition 1####### accepts all signals that start with a 1, whereas 1##01### accepts only a subset of those signals: those that start with a 1 but also have a 01 in positions 4 and 5. The first condition is more general than the second condition because, from an information-theoretic point of view, the condition 1##00### uses more information than the condition 1#######. Note also that the signals accepted by 1##01### are exactly the signals defined by the *intersection* of the sets 1####### and ###01###. (For another example, see figure 6.3.)

The object now is to examine how crossover produces rules that work together at different levels of generality. The rules employed by the situated agents of section 3.2 provide a good starting point. In that simple example, eight detectors react to the following properties:

<moving, airborne, black, large, winged, fuzzy, multi-legged, long>.

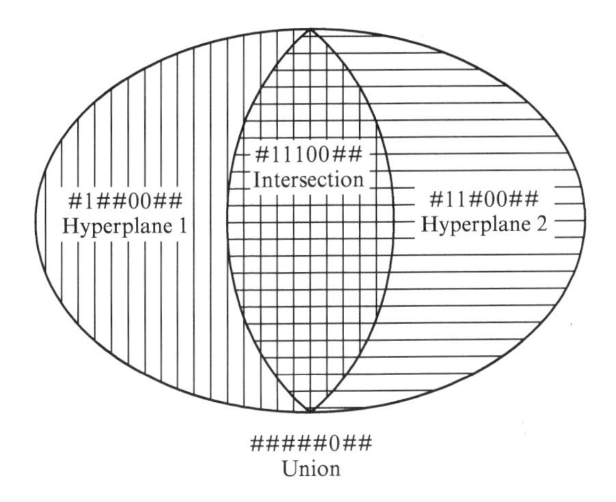

Figure 6.3
Hyperplane (schema) intersection.

A detector a signal with a 1 at position 1 signals "moving," while a 0 at that position signals "not moving," and so on. Accordingly, the condition

1#######

reacts to any *moving* object, whereas the condition

1##01###

reacts only to *moving* objects that are also *small* (not large), and *winged*.

Note that a "moving, small, winged object" satisfies *both* the general condition 1####### and the more special condition 1##01###. Consider, then, an agent that uses the conditions 1######## and 1##01### for rules that cause different actions:

1####### => FLEE

1##01### => APPROACH.

Rather fancifully, we could think of a frog at the edge of a pond, for which most moving objects are predators but moving, small, winged objects are prey. Because both conditions are satisfied by a moving, small, winged object, the signal gives rise to conflicting actions. How is this conflict to be resolved? One way to resolve it is to favor the rule that uses more information (the more specific description). In the above situation, the APPROACH rule would be favored when the moving object was also small and winged; otherwise the FLEE rule would be executed.

With this in mind, it is possible to build a hierarchy of rules, called a *default hierarchy*, in which general rules cover the most common situations and more specific rules cover exceptions (Holland, Holyoak, Nisbett, and Thagard 1986). Default hierarchies offer several advantages to systems that learn or adapt:

- A default hierarchy has many fewer rules than a set of rules in which each rule is designed to respond to a fully specified situation.
- A higher-level rule (a rule with many #'s) is easier to discover (because there are fewer alternatives) and it is typically tested more often (because the rule's condition is more frequently satisfied—e.g., moving objects are encountered more frequently than moving, small, winged objects).
- The hierarchy can be developed level by level as experience accumulates.

With this in mind, interest centers on the way crossover produces new candidates for the default hierarchy.

To begin, cross the general condition 1####### ("moving object") with the fully specific condition 11100010 ("moving, airborne, black, small, winged, fuzzy, multi-legged, long object"). A crossover point between the third and fourth loci yields two new conditions: 1##01010 and 111#####. Both new conditions are more specific than the "general" parent, 1#######, and less specific than the "specific" parent, 11100010. Crossing thus specializes the default rule, 1#######, by using some information from the highly specific condition (which responds to one particular input signal). The result is two new candidate exception rules that become hypotheses to be tested for their contribution to the growing default hierarchy. They are plausible hypotheses because they are based on information already incorporated in the rule system.

It is all very well to say that crossing over *at the right point* produces interesting new conditions for the default hierarchy, but how does this work out when the crossover point is chosen at random? How often will crossover produce interesting recombinations then? Here we return to the notion of linkage. The closer alleles are to one another, the less often they will be separated by crossover. In the above example, if the alleles "moving" and "not airborne" are present in one parent, they will appear together in seven out of eight offspring. In effect, loci that are close together tend to be tested together. They become de facto building blocks for the default hierarchy, acting much like a single more complex specification. For strings that are of a more realistic length for signals and condition (say, 100 loci), alleles that are within an interval of ten adjacent loci will be passed on as a group nine times out ten. A set of adjacent alleles often comes to designate a tag, whereupon recombination readily modifies default hierarchies and other signal/boundary systems by putting the tags in new contexts.

What happens when alleles that could designate a useful building block are widely separated? Because they are widely separated, they would rarely be passed as a set to succeeding generations, even if they tested well in the current generation. There is a genetic operator, called *inversion,* that rearranges alleles on a string, so that different sets of alleles are close together in different subpopulations (Dobzhansky 1977). Under inversion, differing arrangements of alleles can be tested and passed on, regardless of the initial sequencing of the genes.

The persistence of a set of alleles under selection determines its contribution to coevolution within a signal/boundary system. Determining persistence is an easy problem if each allele makes a contribution that doesn't depend on the context provided by other alleles. There are standard procedures (Fisher 1930) for assigning a performance number (fitness) to each allele so that the total contribution of all the alleles is simply the *sum* of those numbers. However, in almost all realistic signal/boundary systems alleles *do* interact. For instance, changing a single allele in a tag can cause large changes in the destination of a signal carrying that tag, with corresponding changes in the signal/boundary network. Furthermore, as a signal/boundary system evolves, tags begin to coevolve and interaction increases. The result is a progressive elaboration of hierarchies, "production lines," and the other complex interactions discussed earlier. With this in mind, the emphasis from this point on will be on models and theories that help us understand coevolution's non-additive contributions to complex interactions.

6.2 Reproduction of rules and agents

Recombination is a strong method for introducing plausible new rules, but those rules are hypotheses that must be tested and selected. Darwinian selection has been suggested, but how do we implement it in this rule-based approach? There are three requirements:

- There must be a *population* so that the rules can compete.
- There must be a *comparison* of rules that results in a rating that indicates the rule's performance within the population.
- There must be a *selection* process, so that rules that are successful in the competition can be favored.

The first requirement, a *population,* is satisfied by classifier systems because they involve many rules that can interact simultaneously without concern for logical conflicts.

The second requirement, *comparison,* requires a process for measuring a rule's performance within a population. In the realm of computer-based models, this process goes by the name *credit assignment* (Lanzi 2000). A rule, for example, can be assigned a *strength* that reflects its level of performance. If the rule is treated as a hypothesis, this strength can be thought of as its level of confirmation. Because many rules can be satisfied at one time, there can be a competition among rules for activation, stronger rules being favored. To favor the formation of default hierarchies, the credit assignment algorithm can adjusted so that more specific rules (rules with fewer #'s) gain or lose strength more rapidly. More specific rules that are confirmed become stronger and override their less specific competitors.

The strength of a rule enters directly into meeting the third requirement, *selection.* The rule's strength can be used as its Darwinian *fitness.* Rules that are stronger (better confirmed as hypotheses) should contribute more to the evolving organization of the signal/boundary system. A straightforward way to accomplish fitness-biased evolution is to cross-breed strong rules, using the offspring as new hypotheses. Cross-breeding accomplishes the recombination of building blocks that provides plausible new hypotheses with the mixture of specificity that encourages parsimonious default hierarchies. The new hypotheses replace weaker rules in the population, not the parents.

To understand the spread of building blocks under this regime, we need a generalization of Fisher's fundamental theorem about the spread of individual alleles. One such generalization is called the *schema theorem.* (See Mitchell 1996.) A *schema* is an arbitrary allele cluster, specified using the symbol * to designate places along the chromosome not belonging to the cluster. As an example, the cluster consisting of allele 1 at position 2, allele 0 at position 4, and allele 0 at position 5 is designated by the string *1*00** . . . *.

In the population at time t, schema s will appear in a certain number of chromosomes, N(s,t), possibly 0. The schema theorem specifies the (expected) number N(s, t + 1) of chromosomes carrying schema s in the next generation t + 1. A simplified version of the schema theorem has the following form:

$$N(s, t + 1) = u(s,t)[1 - e]N(s,t),$$

where u(s,t) is the average fitness of the chromosomes carrying schema s at time t (the *observed* average fitness) and where e is the overall probability (usually quite small) that the cluster s will be destroyed (or created) by mutation or crossover. Note that e does become large if the alleles in the cluster are spread over a considerable part of the chromosome (low linkage). This formula for N(s, t + 1) can be restated in terms of probabilities (proportions) P(s,t), a form typical for mathematical genetics, by noting that P(s,t) = N(s,t)/N(t), where N(t) is the size of the population at time t. If the size of the population is fixed at N—say, if each pair of offspring replaces a pair of strings already in the population—then P(s,t) = N(s,t)/N and the schema theorem has the form

$$P(s, t + 1) = u(s,t)[1 - e]P(s,t).$$

In words, when e is small the probability of finding a schema s in the next generation increases or decreases approximately according to the average fitness of its carriers in the current generation.

The schema theorem shows that *every closely linked cluster of alleles* present in the population increases or decreases its proportion in the population at a rate largely determined by the average fitness of its carriers. In particular, schemata consistently associated with above-average strings spread rapidly through the population. As a result, crossover regularly tries out new combinations of these above-average schemata, treating them as building blocks for further attempts at improvement. Though a genetic algorithm directly processes only N(t) strings in a generation, it effectively processes the much larger number of schemata carried in the population. For example, in processing a population of 100 strings of length 40, the genetic algorithm tests on the order of 2 exp(40) ~ 16,000 schemata (a multiple of the number of schemata carried by a single string of length 40). For problems in which schemata capture regularities in the search space, this is a tremendous speedup.

Because rules, as well as their parts, interact to preserve an agent in a signal/boundary system, it is necessary to define the overall fitness of the *set* of rules defining the agent. One way to accomplish this is to return to the discussion rules in relation to reactions. Under this view, rules are fashioned from a set of elements (the letters) that are conserved. Conservation of elements requires that an agent can replicate only by collecting elements through its interactions with other agents and with non-agent parts of its environment. That is, replication isn't simply a matter of writing down the new rule; the agent must acquire the letters required to write the rule. The agent can, for example, use a *reservoir* to collect the elements provided by "dismantling" the signals (resources) it acquires through interactions. When the reservoir contains enough elements to make copies of all the agent's rules, the agent replicates. This process mimics, in a simple way, a cell's use of resources to make copies of its proteins, DNA, RNA, etc.

6.3 Stage setting and fitness

At this point we can tie the strengths assigned to individual rules under credit assignment to the overall fitness of the set of rules defining the agent. Each time an agent acquires some useful elements in its reservoir, the rules active at that time automatically have their strengths augmented in a kind of direct reinforcement or conditioning. This still leaves the problem of setting the strengths of "stage-setting" rules—that is, rules not directly involved in element acquisition.

As an analogy, think of payments to early members of a chain of "middlemen" stretching from ore mining, through smelting and part production, to automobile assembly and the ultimate consumer. Clearly payments by the ultimate consumer don't go directly to the ore miners. Instead, the payments are passed up the chain of middlemen through repeated buying and selling. A similar procedure can be used to reward stage-setting rules by having rules buy and sell the signals they process. Each rule satisfied by a signal bids a proportion of its current strength (its "cash in hand") to acquire the right to process the signal. High-bidding rules pay their bids to the sender(s) of the signal(s) that satisfied them (see figure 6.4) and place their signals on the signal list. In effect, each winning rule has bought the right to

A rule bids 20% of it strength.
A rule active when a reservoir is incremented receives a payoff of 100.

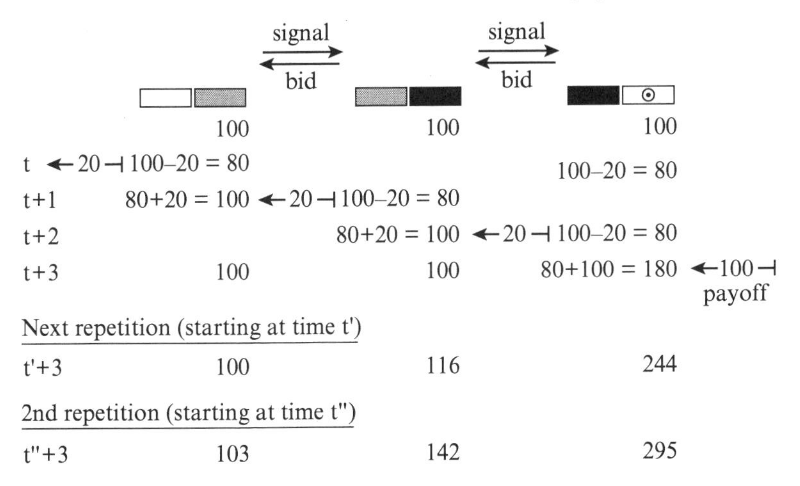

Figure 6.4
Credit assignment by bucket brigade.

"sell" its signal to other rules; it then receives payments in turn from the next set of active rules that are satisfied by its signal. In this way, each rule acts as a "middleman" in the market for signals. Rules that receive more in payments than they pay out in bids automatically become stronger—they make a profit.

Consider, then, a chain of rules leading to an "ultimate consumer" rule that fills the agent's element reservoir. Because a rule that fills the reservoir has its strength immediately increased, it will make a higher bid next time it is activated. So the next time, it pays more to its "suppliers"—the rules that send signals to it. Using standard economic theory, then, it isn't difficult to show that rules belonging to a chain leading to the "ultimate consumer" rule become stronger over time (Samuelson and

Nordhaus 2009). In particular, rules that act early, setting the stage for the later reservoir-filling action, become stronger.

In this way strength and fitness are tied to each other because chains that lead to reservoir filling actions are strengthened and play bigger roles in determining the agent's behavior. Successful chains fill the reservoirs more rapidly, allowing the agent to replicate more rapidly. In Darwinian terms, such an agent is more fit and contributes more to the evolution of the signal/boundary system. Note that fitness, defined in this way, doesn't require an explicit fitness function that assigns a fitness number to each agent. The fitness is implicitly defined, depending on both the agent's rules and the availability of the elements it needs for replication.

Summary

The framework developed to this point centers on agents defined by signal-processing rules. Recombination and reproduction are then applied to the rules to explore further possibilities (hypotheses). An exchange of signals between agents amounts to an exchange of resources when the rules are looked upon as reactions, as in chapter 3. Signals, guided by tags, also coordinate the agent's interior workings. When tags are looked upon as schemata that can be recombined and reproduced we have a direct way of studying coevolution in signal/boundary systems. Our next step is to examine more precisely the effects of semipermeable membranes on this coevolution.

7 Urn Models of Boundaries

Generated reaction networks, "glued together" by means of tags, capture a substantial portion of the desiderata for an overarching framework—a grammar, a geometry, programmability, and reproduction. Additionally, because the tags can be recombined to modify network structure, it is straightforward to incorporate mechanisms for adaptation and evolution of the network. There is reason to believe, then, that generated reaction networks would fit all four requirements *if* there were a simple way to use tags to implement *boundary-induced constraints* on resource and signal flow. That is the concern of this chapter.

7.1 Urn models

Boundaries filter the flow of resources through a reaction network. In section 3.4 these flows were discussed in terms of elastic collisions, a so-called billiard-ball mechanics. This chapter uses the urn models of elementary probability to capture the basics of billiard-ball mechanics in a way that allows filtering. (Feller's classic 1968 text provides interesting examples of urn models.) The present section shows how urn models can capture particle flows; the next section describes a way of augmenting urn models with tags to capture membrane-like constraints on the flows.

In the simplest urn models, we place a mixture of balls of different colors in an urn. The proportion of each color determines the probability that a given color will appear in a random draw from the urn. For example, if there are 20 black balls and 10 white balls in the urn, the probability of drawing a white ball is 10/(10 + 20) = 1/3. More generally, if there are n_1 black balls and n_2 white balls, the probability of drawing a white ball is

$$p_2 = n_2 / (n_1 + n_2).$$

Still more generally, if there are k colors with n_i balls of color i in the urn, a random draw will produce a ball of color i with probability

$$p_i = n_i / (n_1 + n_2 + \cdots + n_i + \cdots + n_k).$$

How does this apply to a reaction network satisfying billiard-ball mechanics? If a different ball color is associated with each reactant, then p_i is the concentration of reactant i. Accordingly, the probability that two reactants, i and j, will collide under billiard-ball mechanics is simply approximated by two successive draws from the urn. Once the reactants collide, the probability that i and j will react to produce balls of colors k and m is given by the reaction rate $r_{ij|km}$. That is, we draw two balls i and j from the urn; then, with probability $r_{ij|km}$, we replace them with balls of colors k and m.

A given collision may result in different products. For example, the collision of i and j might result in products k and m or products g and h. This can be handled by conditional probabilities, where $r_{ij|km}$ is treated as the conditional probability that products k and m are formed when i and j collide. For each pair i and j, there will be a set of conditional probabilities

$R_{ij} = \{r_{ij|km}, r_{ij|gh}, \ldots\}$

indicating the probabilities of different possible products. This includes the conditional probability $r_{ij|ij}$ that i and j are unchanged (that is, that there are no reaction results from the collision). The probabilities in each R_{ij} must sum up to 1 to indicate that one of the possibilities, including no reaction, must happen. If only one result is possible—say, if i and j always react to produce k and m—then the only entry in R_{ij} is $r_{ij|km} = 1$.

When a new reactant (a new color)—say, x—enters the urn as the result of a reaction, new reactions become possible. We then face the problem of determining reaction rates $r_{xi|uv}$ for reactions involving the new reactant. We could list all possible reactions and reaction rates at the outset—a daunting task if the set of possible reactants is large (as with the set of proteins). Compiling such a list would be as impractical as trying to list the fitnesses of all possible organisms in different groupings. The alternative, more in keeping with our objectives, is to *generate* reaction rates from the reactant's tags, treating the tags in the same manner as the active sites of proteins. To set up this rate-generation procedure, let us combine the concept of *valence* from elementary chemistry with the tag concept.

Diffusion

The first step before melding valence and tags is to expand the urn models to include multiple urns, with balls being drawn from one urn sometimes being placed in another. The result is a "flow" of balls into and out of urns. If we think of the urns as placed in varying locations, we bring into play the second of the main requirements for signal/boundary theory: spatial separation. The resulting spatial flow is the counterpart of diffusion in a mixture of chemicals.

To mimic diffusion with urns, then, let each urn correspond to some region in a "vat" of reactions. The proportion p_i of balls of color i in a "regional" urn is proportional to the concentration of reactant i in that region. By drawing a ball from an urn and placing it in a randomly chosen adjacent urn, we mimic random particle diffusion, the kind of diffusion brought about by random, color-independent, elastic collisions between the balls.

As a mental test, think of all the black balls of one color clustered in one region of the "vat." Intuition tells us that, as the black balls move and collide at random, they should spread uniformly throughout the vat. How does this effect show up in the urn model? At the beginning, the balls will all be clustered in a single urn. If we try to draw a ball from an empty urn, there will be nothing to draw, but a draw from the urn with all the balls will move a ball to an adjacent empty urn. Over time, repeated draws will spread the balls throughout the urns. We can improve upon this model by simply drawing at random from the whole set of balls, regardless of location, so that we waste no time trying to draw from empty urns. This has the desired effect that most of the "action" will center on urns with many balls. Stated another way, diffusion occurs most rapidly from regions that are crowded. This version matches both intuition and experiment.

To go further, consider again the urn with 20 black balls and 10 white balls. Let there be a second urn with 10 black balls and 20 white balls. Pick a ball at random from the first urn and place it in the second urn. The odds are 2 to 1 that the ball chosen will be black. If a black ball is chosen, the number of black balls in the second urn will be increased. We can also draw a ball at random from the second urn and place it in the first urn, in which case it is probable that the number of white balls in the first urn will be increased. What happens if we make a succession of choices, alternating between the first and the second urn, always moving the ball drawn to the other urn? It is easy to convince yourself that after a while the two urns will tend to have equal numbers of white and black balls. That is, on average, each urn will have 15 white balls and 15 black balls. This is similar to the effect of random mixing, or diffusion, where we expect the average number of each kind of particle to be uniformly distributed.

What happens if balls of one color (say, red) are distributed uniformly over all the urns, except for one urn that has a large number of white balls and no red balls? Over time, how will the red balls come to be distributed? Some thought shows that red balls will begin to appear in the urn full of white balls even though it is "crowded." Indeed, they will appear there just as rapidly as when the urn is empty. Again this outcome matches experiment, though intuition might suggest otherwise. There is no "crowding effect." Each ball moves through the urns in a kind of random walk, as if there were no other balls around. So it is with each distinct reactant in a vat.

7.2 Urns with entry and exit conditions

Recent studies (e.g., Slack 2009) put more and more emphasis on receptors and signals in maintaining compartmentalization. Semi-permeable membranes, with either gates or pumps, are a major observable counterpart of this compartmentalization. They can be simulated by adding entry and exit conditions to the urns introduced in the preceding section. The basic idea is simple. A ball can be drawn from (placed in) an urn only if its "color" satisfies one of the urn's exit (entry) conditions. To use our earlier ideas about conditions and tags, particle "colors" are replaced with tags, with a distinct tag for each color. With this provision, the urns are assigned entry and exit conditions defined by strings over {1,0,#} as for classifier rules. Urns augmented with entry and exit conditions provide a representation of membranes and boundaries that allows direct "visual" representation of boundaries, use of collision and reaction probabilities to study effects of boundaries on flows between urns, and provision for evolution of boundaries though modification of urn conditions. Let's look at this approach more carefully.

Constrained diffusion

Though the provision of "rules" determining which "colors" of balls are allowed to enter or exit an urn takes us beyond the usual urn models of probability theory, it retains the possibilities of mathematical analysis. In this model, each ball is assigned a tag or multiple tags (think of colors again, but now we allow the ball to have "stripes"), and each urn is assigned a set of entry and exit conditions. Only a ball with a tag that satisfies an urn's *entry condition* can enter that urn. That is, a randomly selected ball will be able to enter a randomly selected urn only if (one of) its tag(s) matches one of the urn's entry conditions. If there is no match, the ball is "turned aside" and returned to the urn from whence it came. The same format is used for *exit conditions*—when a ball is drawn for diffusion, it can leave the urn only if one of its tags meets one of the exit conditions; otherwise it simply remains in place.

By treating entry and exit conditions as classifier-rule conditions, we arrive at a rule-based urn model. The "entry classifier" accepts a reactant (ball) that carries an appropriate tag for the entry condition, and "passes it through" to the interior (recall the *pass through* symbol in section 3.3). The exit classifier acts similarly. With this addition, both boundaries and reactions can be specified using classifier rules. This uniform treatment provides a way of relating the evolution of signals to the evolution of boundaries. (See chapter 14.)

The constrained diffusion provided by entry and exit conditions provides substantial local departures from uniform mixing. That is, certain regions in the space will have higher than average concentrations of selected reactants. Going back to the reaction equation (chapter 3), we see that some reactions will proceed at a higher rate because of these increased concentrations. Thus, just as membranes allow cells to favor certain reactions, this broadened system of urns and rules favors selected signal/boundary interactions. This enhancement, in combina-

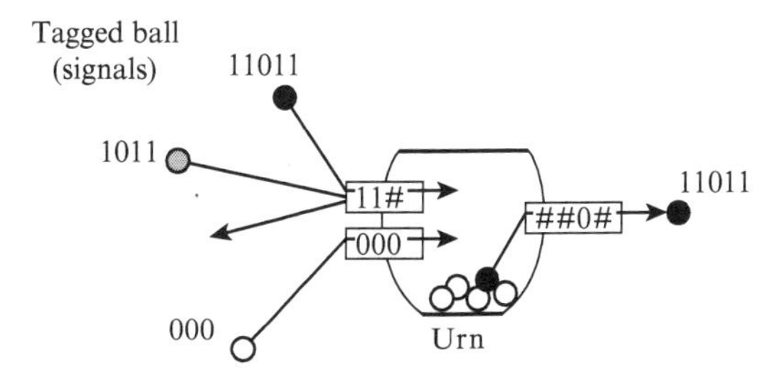

Figure 7.1
A tagged urn.

tion with feedback, recirculation, and catalytic effects, gives evolution a powerful way of favoring reactions that increase fitness. We will examine these evolutionary possibilities in several contexts.

In particular, constrained diffusion provides an opportunity for "production lines" and "cascades." A case in point is Adam Smith's example of specialists in a production line making a previously rare item widely available. Similarly, biological cells use semi-permeable membranes for specialization and cascades, so that reactants that were rare can become common. Urns with entry and exit conditions expand these possibilities to general signal/boundary systems, with one urn for each stage in a coordinated line of agents. The exit condition of each urn in the line corresponds to the entry condition of the next urn in the line. Each urn can favor certain interactions, in effect becoming a specialist for that interaction, and output products at each stage are tagged so that they are accepted by succeeding specialist urns.

This potential for sequencing also offers program-like possibilities. The set of reactants present in an urn becomes analogous to the content of registers in a computer. In a programmed computer, the content of a register can be treated either as data or as an instruction, depending up how it is accessed. When a register's content is interpreted as an instruction, a part of the string (much like a tag) is read as the "location" of the input data for the instruction. A similar description holds for reaction nets. A protein that acts as a catalyst serves as an instruction, and the catalyst's conditions determine the proteins that serve as its data. This comparison to computer instructions is more than a metaphor. As we'll see, urn models, like classifier systems, are "computationally complete." An appropriate urn model can carry out any process that can be defined by a computer program. Indeed, because many reactions (instructions) can be executed simultaneously, the urn model acts a parallel computer.

7.3 Building agents with urns

To fit this rule-based urn system within the framework of complex adaptive systems, we must define *cas* agents in terms of urn-contained reactions. Defining agents in this way is straightforward except for the requirement that an agent be able to reproduce. In the present context, reproduction requires that there be a way to make copies of urns.

Because a tagged urn's boundary-like functions are set by its entry and exit tags, we can reproduce the urn by making copies of its tags. To carry out reproduction in the reaction format, the urn-defined agent must collect the generators (from reactants) necessary to make copies of these tags. This process involves two steps:

(1) Collect the required generators through diffusion and agent interaction.

(2) When sufficient generators have been accumulated, make copies of the tags and organize them to form a new copy of the agent.

Step 1 must be implemented in a way that allows for competition and trading between agents. Step 2 could be implemented in terms of reactions; here, however, for simplicity, storage and copying will be presented as pre-defined operations applied directly to the accumulated generators.

An agent that fits the Echo format provides a simple starting point for reproducing urn-defined agents. The simplest Echo-like agent is defined by an *offense* tag, a *defense* tag, and a *reservoir*. The reservoir is used to store acquired reactants until there are sufficient numbers of each letter type to make copies of the offense and defense tags. When two agents (call them agent x and agent y) come into contact, there can be an exchange of reactants. Such an exchange is mediated by the offense and defense tags. The offense tag of agent x is matched against the defense tag of agent y to determine which reactants move from agent y to agent x. (See figure 7.2.) If the match is poor, agent x gets no reactants from agent y—agent y has a "good defense" against agent x. A better match moves some reactants from agent y's reservoir to agent x's reservoir. A close match causes agent y's reservoir to be emptied into agent x's reservoir,

The offense tag ▯ of agent I is matched against the defence tag ▯ of agent II, locus by locus, and vice versa.

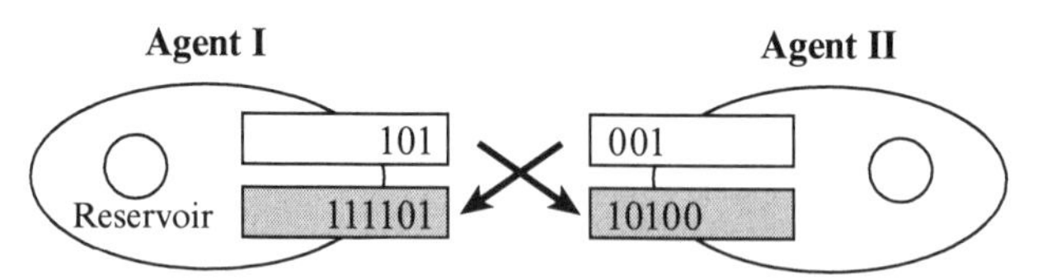

Match at locus = +2
Mismatch at locus = –2
Each defence locus beyond the end of the offense tag costs –1

In the example above:
Agent I scores 2+2+2–1–1 = +4
Agent II scores –2–2+2–1–1–1 = –5

The higher the score, the more resources are transferred from the other agent, up to and including the numbers in the tag (the "death" of the other agent).

Figure 7.2
Echo offense and defense tags.

and agent y's tags are dismantled and transferred—agent x "kills" agent y. The same procedure is simultaneously carried out for agent y. That agent's offense is pitted against agent x's defense. Notice that it is possible that the agents may simultaneously kill each other. On the other hand, a modest symmetric match between agents causes the agents to a "trade" the surplus reactants in their reservoirs. This simple scenario of matching offense and defense can be carried much further. For example, as will be detailed in section 13.2, one agent may engulf another when the match scores are badly asymmetric.

If we think of an urn's entry and exit tags as analogous to offense and defense tags, and of the urn's interior as a kind of reservoir, then we can treat urn reproduction as analogous to the reproduction of Echo agents. There are, of course, some important details to be filled in. Most important, when the agent is composed of many urns, the reactants held in the "reservoirs" of the component urns should be shared and the reproduction of the component urns should be coordinated to produce a complete offspring agent. Beneath these two requirements is a requirement that urns be able to "contain" other urns in order to allow for the full range of signal/boundary interactions. At this point, I will only sketch a way of meeting these requirements. Hierarchical arrangements and reproduction of agents with interior structure will be discussed in detail in the next chapter.

The main point here is that hierarchical arrangements of urns can be achieved by coordinated assignment of urn entry and exit tags. The discussion of membrane hierarchies in section 3.4 outlined a way of assigning tags to attain a hierarchy of enclosures. By adapting this procedure to the entry and exit tags for urns, it is possible to set up a hierarchy of urns, even though the urns "stand side by side." That is, no urn actually contains another urn, even though the tags enforce a hierarchy. And a much simpler mathematical approach in which an urn contains only balls, not other urns, is possible; that will be discussed in the next chapter.

The simplest model for urn reproduction, then, uses a meta-operation that automatically reproduces an urn when it has accumulated sufficient reactants to make copies of its tags. In a more complicated model, this meta-operation can also be defined via coupled reactions, as will be outlined in chapter 14. When several urns are used to define an agent, the process is easier to understand if all the component urns are reproduced simultaneously. In this case, when a component urn accumulates enough resources to reproduce its tags, it sends a "ready" signal to a meta-operator. The meta-operator collects the "ready" signals from all the component urns until all the components indicate readiness for reproduction. Then the reproduction operator is activated. Note that the list of urns used by the meta-operator defines the reproducing agent. Even this more sophisticated operator can be defined in terms of coupled reactions; for an example, see chapter 13.

7.4 Putting it together

Taken together, the pieces of apparatus collected to this point offer a way of meeting the requirements for a signal/boundary theory set forth in the opening section: (i) a grammar, (ii) a geometry, (iii) programmable agents, and (iv) reproduction by resource collection. Signal-processing rules provide the starting point by offering a way of using a simple grammar, with a limited alphabet, to generate complex signaling networks, meeting requirement i. *Tags* for addressing are a special part of the grammar, providing natural building blocks for modifying the networks via recombination. In particular, some of the tags can be used to provide a coordinate-like separation between urns. This coordinate-like use gives us a direct way of providing a geometry within the grammar, meeting requirement ii.

The *cas* framework can be then brought into play by using signal-processing rules to define the activities of *agents*. In particular, agents are defined by using tags to collect a set of signal-processing rules into an interactive bundle, using urn entry and exit conditions to define the agent's outer boundary. Requirement iii is met by using *classifier rules* to implement signal processing, thus realizing computational completeness. Then, by using *reactions* defined by classifier rules as the agent's mode of signal processing, it is possible to introduce the mathematics of particle (billiard-ball) mechanics without losing computational completeness. As a result, the familiar equations relating concentrations and reaction rates become applicable to general signal/boundary dynamics. Urns with entry and exit tags extend the reaction formalism to semi-permeable membranes, using tags and parts of tags as building blocks for the membranes. In addition, tagged urns placed in an underlying geometry directly introduce the spatial effects of membrane-constrained diffusion, thus reinforcing requirement ii. (In chapter 15, tagged urns provide a way of introducing an important piece of mathematics, Markov processes, for describing whole signal/boundary systems.)

This framework, centered on urn models of *cas* agents, provides an approach to a basic question: How do signals and boundaries coevolve? To answer this question, reproduction with variation and selection must be added to the framework, the implementation of requirement iv. The story line is Darwinian. The properties of an urn, and the balls (reactants) it contains, are determined by tags. Because of the conservation-of-elements requirement for reactions, the elements (generators) used to copy tags cannot be created; they can only be collected. Thus, a boundary can be copied only if sufficient elements are collected to make a copy of the corresponding entry and exit tags, and an agent's rate of reproduction is determined by its ability to collect relevant elements. Recombination (chapter 6) ensures that the building blocks of the agent's critical processes propagate to future variants. The object now is to see if this framework can help us understand the boundary hierarchies prevalent in signal/boundary systems.

8 Boundary Hierarchies

8.1 Hierarchical organization of urns

As we have seen, hierarchies and the signaling networks that tie them together are a pervasive feature of all complex adaptive systems. It is natural to ask why this is so. In the present context, this becomes a question about the ways in which the coevolution of signals and boundaries favors hierarchical organization. In particular, does the recombination of entry and exit rules for urns produce hierarchies?

The obvious way to provide a hierarchical organization of urns is to place urns within urns. However, this way of attaining a hierarchy complicates both the presentation and the analysis. Instead, there is an alternative approach, more friendly to analysis, that can be achieved by an appropriate use of entry and exit tags on urns that stand side by side. Now is the time to examine this alternative in more detail.

When one semi-permeable membrane encloses another, only reactants that pass through the outer membrane can reach the inner membrane. To achieve this effect with entry and exit tags on urns, the trick is let the urn corresponding to the interior membrane have an entry tag that is satisfied only by balls drawn from the containing urn—the "interior" urn has an "address" that can be satisfied only by balls drawn from the "exterior" urn. With this approach, we can study boundary hierarchies using the same kind of analysis that applies to diffusion between urns.

By giving the corresponding addresses a hierarchical organization, this side-by-side approach can easily be extended to mimic a succession of enclosures. The addresses are organized in a way that mimics a formal report, with section 1 containing sections 1.1 and 1.2, section 1.2 containing subsections 1.2.1, 1.2.2, and 1.2.3, and so on. Correspondingly, an urn with tag 1.2 can accept only balls drawn from "enclosing" urn 1, and so on. This addressing procedure has two advantages when tagged urns are used to study signal/boundary systems:

(i) The hierarchy of enclosures can be changed by simply modifying tags. It is easy to provide the tags with building blocks that mimic the organization (section, subsection, sub-subsection) of formal reports. Such building blocks offer interesting possibilities for adaptive mechanisms relevant to the origin and evolution of signal/boundary hierarchies.

(ii) Markov processes, a well-studied mathematical formalism, can easily be extended to study these tagged urns. This formalism, which will be examined more closely in chapter 15, provides a foundation for studying boundary hierarchies with the help of mathematical modeling and analysis.

An increasingly complex hierarchy makes possible ever greater control, with a corresponding increase in adaptive potential. By enhancing progressive separation of a uniformly

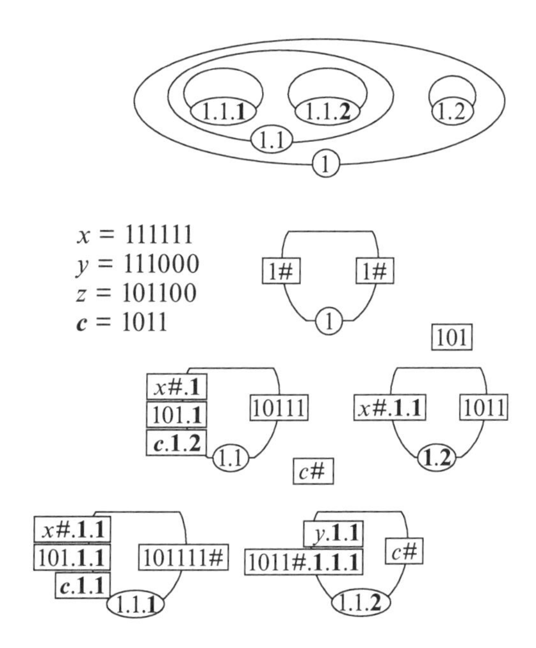

The **suffix** on each urn **entry** tag specifies the urn(s) from which incoming balls may be drawn.

Figure 8.1
An urn hierarchy.

mixed set of reactants, hierarchical enclosures can direct particular reactants to appropriate "specialist" reactions. The specialists can then direct their products to other specialists at the same level, or they can direct them to an interior specialist for further refinement. Cross-breeding between "layered address tags" can place pieces of extant hierarchies in the context of other enclosures. Specialist urns become building blocks for new "production lines" and new hierarchies, generating new varieties without resorting to random changes. This topic will be investigated at some length in chapter 14.

One broad conjecture about hierarchies of enclosure is that they increase the rate of production of selectively advantageous reactants. Though it is relatively easy to establish the efficacy of a particular signal/boundary hierarchy, it is another matter to formulate testable conjectures about the evolution of such hierarchies. The conjecture must not contradict what is known about different signal/boundary hierarchies, and it must also be formulated in a way that makes its consequences testable. Formal computer-executable models offer a considerable advantage in meeting this objective because they make premises explicit that are usually hidden in a verbal argument. If a conjecture cannot be made to work when it is implemented as a computer-executable model, its interest as an explanation is greatly limited. Even when the execution of the model fits what is known, that makes the model only a plausible conjecture—its executable consistency says only that a signal/boundary hierarchy *could* arise in that way. That is still an advance in signal/boundary studies, where even vague conjectures are hard to come by.

Urns with entry and exit tags and tag-directed signals will soon be examined as a way of generating computer-executable signal/boundary conjectures, but we still can go a certain distance using the more familiar equation-based models.

8.2 Coupled interactions—a first look

The following sequence gives a simplified version of Adam Smith's example of actions performed in sequence to achieve a desired end product (a straight pin):

molten metal =[1]=> wire =[2]=> wire with head =[3]=> straight pin.

Here each number refers to a distinct activity: [1] draw metal, [2] clip wire and form head, [3] sharpen. At each stage of the process, signals (light rays, odors, verbal cues, and so on) cue the artisan to carry out the appropriate activity. For example, the characteristics of the molten-metal stage—red, fluid, hot, etc.—cue the artisan to execute wire-drawing. The product of each stage is then passed on to the next stage for further processing. In other words, the processing is executed via a sequence of compartments.

Recasting the production line as a signal-processing system sets it within the *cas* framework, opening it to study by means of classifier rules. As was done earlier, signals coming from the environment can be represented as binary strings, each locus in the string indicating the presence (1) or the absence (0) of a particular property. For example, we can use ten properties to specify the cue signal from the environment:

hot | *fluid* | *bright* | *large* | *long* | *thin* | *flexible* | *cylinder* | *attached* | *disc* | *sharp*.

(Of course a realistic example would involve many more properties.) With this provision, a particular string of ten bits represents a particular the signal from the environment.

If the agent were to use only *hot, fluid,* and *bright* as a condition for initiating the first stage of pin production, the corresponding classifier condition would be 111#######. Accordingly, the first stage,

molten metal (hot, fluid, bright) => draw metal,

would be represented by the classifier rule

111####### => <1>,

with <1> a signal that initiates activity [1]. The subsequent two stages would be represented similarly:

Wire (*not* hot, *not* fluid, *not* bright, long, thin, flexible) => clip wire and form head.

000#1111## => <2>

Wire with head (*not* hot, *not* fluid, *not* bright, *not* long, thin, *not* flexible, cylinder, attached disk, *not* sharp) => sharpen.

0000010110 => <3>.

These three rules could all be executed by a single artisan agent, step by step. Alternatively, each rule could be executed by a separate agent, acting simultaneously with other agents in a production line.

Putting the stages in this rule-based format makes it possible to bring reaction networks into play, using the relation between classifier rules and reactions discussed in chapter 3. That, in turn, provides a well-defined reaction rate for each stage, which makes possible a precise comparison between the production rate of the artisan and production rate of the production line. Looked at as a reaction, the first activity becomes

artisan(u) + molten metal(v) <=> artisan(u) + wire(x)

u + v <=> u + x,

where both a forward and a backward reaction rate, r_1 and r_1', are assigned to the reaction. The forward reaction rate corresponds to the artisan's rate of production of wire, while the backward reaction rate corresponds to imperfections that destroy the wire. Note that the artisan acts as a kind of catalyst, because the artisan is unchanged in the reaction. The second step in the process is coupled to the first by using the output x of the first step as the input to the second step. The second step, then, has the form

artisan(u) + wire(x) <=> artisan(u) + wire with head(y)

u + x <=> u + y,

with forward and backward reaction rates r_2 and r_2'. And so it goes for additional steps.

Now a production rate for each step can be determined using billiard-ball mechanics. Two parameters must be specified in order to proceed: (i) the total number n of collisions per second between reactants at each stage and (ii) the probability, given by the forward reaction rate $r_{cd|ab}$, that a collision between reactants a and b produces the reactant c (and d).

For an urn corresponding to a reaction site, n is the number of draws per second made from the urn. In chemical terms, n is the temperature of the reaction. For the first stage of the pin-production activity, n sets an upper bound on the rate at which the artisan processes molten metal, and so on (with different n's) for the later stages.

In chemical terms, $r_{cd|ab}$ is the proportion of collisions between a and b that result in c, where some collisions my simply be elastic collisions with no reaction. For the artisan, $r_{cd|ab}$ is roughly related to skill. The artisan, when presented with a glob of molten metal—a "collision"—may almost always produce wire, or the artisan may often fail. For a very good artisan, $r_{cd|ab}$ would be close to 1; for a very poor artisan, $r_{cd|ab}$ would be close to 0.

In a reaction such as a + b <=> c + d, the concentrations of the reactants {a, b, c, d} determines the direction and throughput of the reaction. If only reactants a and b are present, the reaction moves entirely in the direction of producing c and d; contrariwise, if only c and d are present, the reaction moves entirely in the direction of producing a and b. In the billiard-ball model, the concentration of a reactant x determines the probability p(x) that a given collision involves reactant x. In a collision between a and b, either a can hit b or b can hit a. Accordingly,

$$p(a)p(b) + p(b)p(a) = 2p(a)p(b)$$

closely approximates the probability of a collision involving both a and b. The earlier discussion of the billiard-ball model (chapter 3) then shows that the rate of the production of c, R(c|a,b), in terms of the concentrations of a and b, is given by

$$R(c|a, b) = n\ r_{cd|ab}\ 2p(a)p(b),$$

where $r_{cd|ab}$ is the forward reaction rate.

So far we have been talking about reactions without catalysts. What is the effect of a catalyst on reaction rates? Here it is useful to refer back to the "notched bowl" example of a catalyst in section 3.4. As in that discussion, let h be the energy required for a and b to react. Now consider a billiard ball x that acts as a catalyst. In typical cases, x acts by turning the reaction into a two-step process, where first one reactant (say, a) collides with x, then later the other (say, b) collides with x. If each of the effective contacts with x requires h/2 (just half the energy of direct interaction between a and b), then a and b will "stick to" (react with) x exponentially more often than they stick to each other. (See the discussion of the Maxwell-Boltzmann distribution in chapter 3.) Note that the total energy required to form the two bonds with x is h = h/2 + h/2. If x releases that energy h to form a bond between a and b, then (i) a and b are no longer bonded to x and (ii) x is then available to catalyze the reaction again with another pair of balls. Because the rate of the two-step reaction with x is effectively determined by h/2 rather than h, the catalyst has effectively put a notch in Asimov's bowl. Though actual catalysts often act in a more sophisticated manner, this simple model does aid the intuition. From a rule-based point of view, the catalyst acts as a two-condition rule for processing signals, a point that will be exploited in chapter 13's ontogeny model.

In studying coupled interactions, then, the objective is to use the production rate R, for both forward and backward reactions, to determine the net throughput of the coupled reactions. The next section will do this for the simplest case, a pair of coupled reactions:

{a + b <=> c + d, c + e <=> f + g}.

The comparison will be between the reactions taking place in a single urn (the case of the generalist artisan) versus each reaction taking place in a different urn, where the output of the first urn is passed to the second urn (a two-step production line).

8.3 Coupled interactions—a precise calculation

For the pair of coupled reactions

{a + b <=> c + d, c + e <=> f + g}

let the products of the first reaction, c and d, both be removed from the site at the same rate, so that d doesn't accumulate. For example, the reaction might be

a + b <=> cd,

so that c cannot be removed without removing d. Under this assumption, the concentration of c is determined by three factors:

(1) The production of c (and d) depends on the forward reaction rate $r_{cd|ab}$ and the concentrations p(a) and p(b) of the precursors, $R(c,d|a,b) = 2n\ r_{cd|ab}\ p(a)p(b)$.

(2) The loss rate of c (and d) through the back-reaction depends on the back-reaction rate $r_{ab|cd}$ and the concentrations of c and d, $D(c) = 2n\ r_{ab|cd}\ p(c)p(d)$.

(3) The loss of c and d through the production of f and g depends on the forward reaction rate $r_{fg|cd}$ and the concentrations of c and e, $R(f,g|c,e) = 2n\ r_{fg|ce}\ p(c)p(e)$.

The rate of throughput clearly depends on the intermediate product c. If there is steady throughput (steady state), then the formation of c is balanced by the loss of c through the reverse reaction plus the loss of c through the formation of f + g:

$$R(c,d|a,b) = D(c) + R(f,g|c,e).$$

Using (1)–(3), we get

$$2n\ r_{cd|ab}\ p(a)p(b) = 2n\ r_{ab|cd}\ p(c)p(d) + 2n\ r_{fg|ce}\ p(c)p(e).$$

Assume that replacement provided by external input keeps the concentrations of a and b fixed at p(a) = p(b) = k1. Then canceling the 2n on each side of the equation and substituting k1 for p(a) and p(b) on the left-hand side yields

$$r_{cd|ab}\ (k1)^2 = r_{ab|cd}\ p(c)p(d) + r_{fg|ce}\ p(c)p(e).$$

When d is used in the forward reaction at the same rate as c (because of a coordinated reaction using d—say, e = cd), the equation becomes

$$r_{cd|ab}\ (k1)^2 = r_{ab|cd}\ p(c)^2 + r_{fg|ce}\ p(c)p(e).$$

If e is supplied from outside the system, as with a and b, the production of f is maximized when p(e) = p(c), Under this arrangement, the equation becomes

$$r_{cd|ab}\,(k1)^2 = r_{ab|cd}\,p(c)^2 + r_{fg|ce}\,p(c)^2.$$

Solving for p(c) yields

$$p(c)^2 = (k1)^2\,r_{cd|ab}\,/(r_{ab|cd} + r_{fg|ce}),$$

$$p(c) = k1[r_{cd|ab}\,/(r_{ab|cd} + r_{fg|ce})]^{1/2}.$$

This general equation for the concentration of c makes possible a precise comparison of output f in the one-urn ("artisan") and two-urn ("production line") cases.

Coupled reaction in a single urn

When the coupled reactions occur in a single urn, all the participating reactants {a,b,c,d,e} are present. If we assume that reactant f is removed as soon as it is formed, their "concentrations" must sum to 1:

$$p(a) + p(b) + p(c) + p(d) + p(e) = 1.$$

Now consider a simple numerical example of the production of the end product f (the throughput) of this single-urn system. Set $r_{ab|cd} = 0$ so that there is no back-reaction (if we hark back to the example of the artisan, the artisan makes no mistakes), and set both $r_{cd|ab} = 1$ and $r_{fg|ce} = 1$ so that every collision results in a reaction (the artisan handles all material presented). Then it is easy to see that setting all the concentrations to 1/5 satisfies the concentration equation. Under these provisions, the rate R(f) of production of f, given in terms of the total number of collisions per second (n), is

$$R(f) = 2nr2p(c)p(e) = 2n/25 = 0.08n.$$

More realistically, consider a case in which the reaction rates are not so simple—say,

$r_{cd|ab} = r_{ab|cd} = r_{fg|ce} = 1/2.$

Using the reaction equation for p(c), we get

$p(c) = k1(1/2)^{1/2} = 0.7k1.$

Then setting p(b) = p(a) to maximize the production of c yields the following concentration equation:

$k1 + k1 + 0.7k1 + 0.7k1 + p(e) = 1.$

Finally, to maximize the production of f in the second reaction, set p(e) = p(c) and sum the components in the concentration equation to get

$4.1k1 = 1,$

which yields

$k1 = 0.244.$

Substituting in the equation for p(c) gives

$p(c) = 0.1725,$

so the rate of production of f is

$R(f)= 2nr2p(c)p(e) = 0.03n.$

It is noteworthy that even a large change in reaction rates (say, from $r_{cd|ab} = r_{fg|ce} = 1$ and $r_{ab|cd} = 0$ to $r_{cd|ab} = r_{fg|ce} = r_{ab|cd} = 1/2$) has little effect on the throughput of the coupled reactions.

Coupled reaction in two urns

Now consider the same coupled reaction taking place in two urns. In this case, urn 1 contains the reactants a and b, and urn 2 contains reactants c and e. As before, a, b, and e are kept in constant supply from outside the system. When reactant c is formed in urn 1 it immediately passes to urn 2; similarly, when f is formed in urn 2 it immediately passes outside the system. As a consequence, in urn 1

$$p(a) + p(b) = 1$$

and in urn 2

$$p(c) + p(e) = 1.$$

When the reaction rates are $r_{cd|ab} = r_{fg|ce} = r_{ab|cd} = 1/2$, and the concentration of e is matched to c,

$$R(f) = 2n\ r_{fg|ce}\ p(c)p(e) = n/4 = 0.25n.$$

Comparison

A comparison of the outputs in the one-urn and two-urn coupled-reaction cases is enlightening. The output of f in the two-urn case, R(f) = 0.25n, is eight times the output in the single-urn case, R(f) = 0.03n. Even if we allow for two artisans working side by side (equal to the two specialists in the two-urn case), the output of the production line is larger by a factor of 4. Just what has happened? The effect of separating the reactions into two compartments is to increase the local concentrations of the reactants relevant to the two reactions. Because the reaction rate increases as a product of the concentrations, the reactions proceed at a much enhanced rate when separated. Roughly, the increased concentrations correspond to the increased attention that individuals on a production line can give to their specific tasks.

The simple lessons provided by this example are widely applicable. The forward reaction rate corresponds to the difficulty of the assembly task—the smaller the rate, the more difficult the task. The backward reaction rate corresponds to the difficulty of maintaining an assembly—the larger the rate, the more difficult it is to maintain the assembly. In chapter 15 this correspondence will be extended to rule-based systems, using the correspondence between rules and catalysts discussed in section 3.4.

8.4 What next?

To transform Adam Smith's example into a model susceptible of calculation, three steps were necessary: representing the production line as a signal-processing system, recasting the signal processing as coupled reactions, and using an urn-based billiard-ball mechanics to calculate the throughput. As we'll see in the next chapter, this use of urns to describe production lines provides a way to study the *evolution* of production lines and hierarchical boundary arrangements.

The specific objective in the next chapter is to link the increased throughput provided by internal compartments to an increased ability of the agent to provide variants of itself. To this end, the agent's internal compartments will be treated as a collection of tag-based urns. This treatment yields the constrained diffusion typical of coupled reactions mediated by semi-permeable membranes. Increased throughput, then, lets the agent collect the resources needed to copy its structures more rapidly. Accordingly, increased throughput is linked to the agent's Darwinian fitness—the agent's contribution to future generations. As a direct consequence, the building blocks that specify boundary arrangements (e.g., tags) propagate through the population in the manner specified by the schema theorem of chapter 6.

9 The Evolution of Niches—A First Look

9.1 The idea of niche

In the study of ecosystems, the concept of niche is usually restricted to a specific kind of agent, typically a particular species. The niche is "powered" by a continuing flow of resources, much like a vortex in a fast-flowing stream. The term is similarly used in phrases like "market niche." However, it is advantageous in a general approach to signal/boundary systems to broaden the interpretation so that it includes conglomerates (for example, an array of diverse, interdependent agents). Then the term *niche* can be used to designate the complex interactions that center on a bromeliad in a rainforest (as described in chapter 1) or a government bureau (such as the Securities and Exchange Commission).

Under this interpretation, niche designates a tangle of local interactions with recirculation, allowing resources to be used over and over again. For the bromeliad niche, carbon can be looked upon as a resource passed from one organism to another with little depletion, allowing diverse individuals to live, interdependently, in the same locale. Cash plays a similar role in an economic niche, and its passage through a chain of buyers and sellers gives rise to the *multiplier effect* (Samuelson and Nordhaus 2009). More generally, in a network, a niche is a *community* having a great number of internal connections but relatively fewer external connections (Newman, Barabasi, and Watts 2006), making it possible for the community to exhibit partially autonomous behavior. In each case, recirculation of resources means that activity in the niche cannot be determined by simply adding up the activities of the different agents occupying the niche.

Experimentalists, understandably, want to deal with uniform conditions; hence the usual restriction in ecology to the use of niche for a particular organism in a well-characterized environment. But such formulations yield little information about general mechanisms for niche formation and change. When broader definitions of niche are used, they are typically qualitative. For example, the *Cambridge Dictionary of Biology* defines niche as "the position . . . of an organism within its community . . . result[ing] from the organism's structural adaptations, physiological responses, and innate or learned behavior" (Walker 1990). Though this definition is suggestive, it would be difficult to use it as a guide for finding the mechanisms that generate niches. In the absence of knowledge of these mechanisms, there is little chance of explaining the emergent properties typically ascribed to niches—crowding, competition, mutualism, and the like. (See Gilbert and Epel 2009.)

This chapter provides a first look at niche-relevant mechanisms by presenting a sequence of simple mechanism-based models that exhibit, in order, niche crowding, competitive exclusion, multiplier effects, and niche invasion.

9.2 Bandits with queues—a niche analog

The study of payoff odds for slot machines, sometimes called "one-armed bandits," goes back to the origins of probability theory. The basic problem is to estimate expected winnings (or losses) over an extended sequence of plays of the machine. The usual way to make this estimate is to make several plays (pulls of the arm), calculate the average return per play, and then use that as an estimate of the payoff rate. Many sophisticated techniques in statistics are based on this simple idea (Feller 1968).

The exposition that follows—a more niche-like version of the estimation problem—starts with a "two-armed bandit"—a slot machine with two arms. Each arm pays with a different probability. For example, suppose that arm I pay \$1 with probability 1/4 and that arm II pay \$1 with probability 1/2. In this case, the player maximizes expected income by playing the arm with the higher probability (arm II) all the time. However, if the probabilities are unknown to the player, a subtle question arises: How should the player allocate plays between the two arms to maximize long-term expected winnings? That is, which "niche" should the player occupy?

One approach is to play both arms a fixed number of times (say, n), then play the arm with the higher observed average payoff ever after. In the example given, if the observed averages are close to the true averages, the player will observe an average payoff per play of $1/4 \times \$1$ from arm I and an average of $1/2 \times \$1$ from arm II. On the basis of an accurate estimate, the player will then play arm II ever after, as in the case of full knowledge. Note, however, that it has "cost" the player to obtain the information—the player has played the less good arm n times, suf-

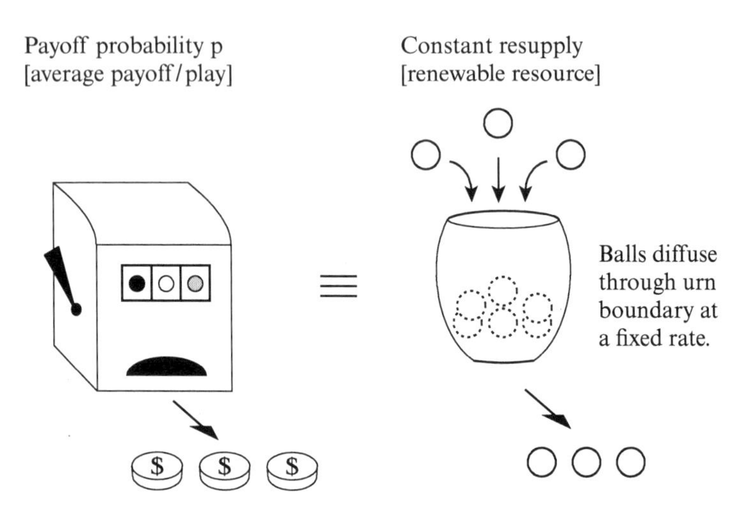

Figure 9.1
A two-armed bandit.

fering a "missed opportunity" cost of \$1(1/2 – 1/4)n = \$n/4. Increasing n increases the accuracy of the estimate, but at increasing cost.

There is another complication. We all know of "runs of bad luck." At Monte Carlo there have been runs of thirty and forty reds in a row on the roulette wheel, even though red and black are equally likely. If, in sampling the two arms, the higher average is the result of one of these "unlucky" runs, then the player will be playing the worse of the two arms ever after. That misestimate leads to an ever increasing decrement relative to what might have been attained. This cost increases without limit as the player continues to play the wrong arm. Is there a strategy that "insures" against this unlucky outcome?

Yes. The player continues to play both arms, but allocates trials at an exponentially increasing rate to the arm that currently has the higher observed average payment (Holland 1992). This strategy has two consequences. (i) The estimates of best and second best become steadily more reliable as additional plays are allocated to each arm. (ii) The exponential increase ensures that, in the long run, the best arm will be played almost all of the time.

The exponential rate of increase of replicating agents suggests a way of implementing this strategy. Treat each of the arms as a niche that supplies resources for an agent's replication. Each time the arm makes a payoff, the agents at that arm replicate. It is easy to show that, as time elapses, the population associated with the more frequently paying arm will increase exponentially relative to the population at the other arm. If the size of the population represents the number of plays of the arm at any stage, the strategy of the preceding paragraph has been implemented. In effect, the population at each arm represents the sampling rate of the arm.

Now, to make the bandits' arms even more niche-like, require that the population of agents queued at each arm divide the payoff equally among themselves. That is, require that they *share* the payoff. For example, let the average payoffs for I and II be $1/4 \times \$1$ and $1/2 \times \$1$, as before, and let there be a total of twelve agents. Consider the case in which all twelve agents are queued behind arm II. Then each of these agents will average $\$1 \times 1/2 \times 1/12 = \$1/24$ per trial. Note that if one of these agents were to move to arm I, forming a queue of length 1, that agent would receive an average income of $\$1 \times 1/4 \times 1 = \$1/4$ per trial, six times as great as the agents queued behind the better arm. Clearly there is an advantage to moving from queue II to queue I in this case, even though the overall input of resources is greater in queue II.

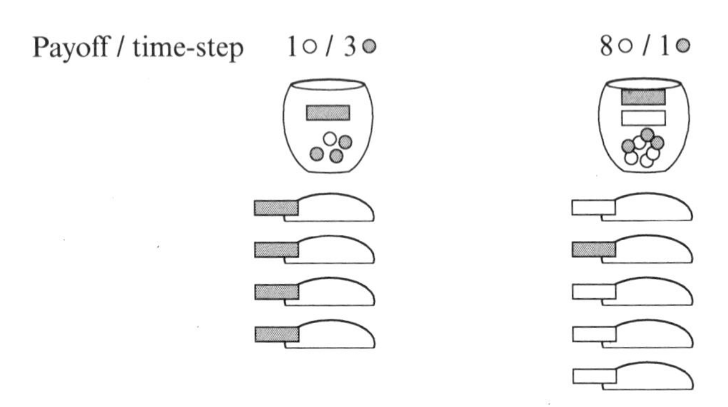

Agents in a queue share resources that match their tag.

payoff ○ / agent □		2.0 ○ / time-step
payoff ● / agent ■	0.75 ● / time-step	1.0 ● / time-step

Figure 9.2
Crowding.

This simple example, then, exhibits a "crowding effect." As the number of players in the queue increases, the payoff *per individual* decreases. The longer the queue, the more "crowded" the niche. A simple calculation shows that players receive equal expected payoffs when each queue length is proportional to the expected payoff. In the present example, having four players in queue I and eight players in queue II yields an equal expected payoff of $1/16 to each player. This model is easily extended to more than two niches by adding arms to the bandit. Each additional arm, then, acts as a different niche, with agents seeking to exploit sparsely occupied niches.

Interestingly, there is a simple algorithm that lets the agents arrive at queues yielding equal expected payoff without explicit agent-to-agent cooperation. Each agent looks to an adjacent queue to see if the average income per agent there is greater than its current average income. If it is, the agent has a fixed probability (say, a flip of a coin) of migrating to the adjacent queue. That is, when the income at the adjacent cure is more favorable, the agent moves if the coin comes up "heads" and stays in place if it comes up "tails." When each agent uses this procedure, the queues' expected lengths rapidly converge to the lengths ensuring that all agents receive the same average payoff, with queue lengths proportional to the payoff rates. Because movement isn't determined by a central executive, this algorithm provides a simple example of distributed control (Han, Li, and Guo 2006).

Another kind of distributed control arises when agent replication based on payoff is used in the queued populations. Here a Darwinian element enters. An agent accumulates payoffs, or resources, until the accumulation reaches a threshold that allows it to reproduce, at which point it adds a copy of itself to the queue. Clearly, agents that accumulate payoff more rapidly reproduce more rapidly. To keep the overall number of agents constant, an agent from a randomly selected queue is deleted each time a replicant is added. The individual queue lengths "settle down" (with modest random fluctuations) when the agents in each queue replicate at the same rate. As before, that rate occurs when each queue length is proportional to its payoff.

The next section builds on these ideas, presenting a sequence of models based on bandits with queues, each queue occupied by diverse agents that both migrate and replicate. By adding additional gizmos to the two-armed bandit, we can proceed to models that offer ever more niche-like characteristics. The models are *not* intended as full-fledged models of niches—they are suggestive, not realistic. They offer a way to build intuition about mechanisms that generate niches. The calculations involved are elementary, but they sometimes require close attention. The reader less interested in rigorous calculation will attain enough comprehension for further reading by jumping to the descriptive conclusions at the end of the calculations for each model.

9.3 A sequence of queued-bandit models

The basic parameters for queued bandits are the following:

R_i is the average amount of resource (payoff) available per time step (play) at queue (arm) i; it may be a vector if resources of several different kinds are available.

L_{ij} is the number of copies of agent type j in queue i. L_i is the total length of queue i = $L_{i1} + L_{i2} + \cdots + L_{ij} + \cdots$.

r_j is the resource amount required for agent type j to reproduce; it may be a vector if agent type j requires a set of distinct resources to reproduce.

d_j is the probability that an agent of type j is randomly deleted. Note that an agent may die of other causes, such as predation.

The basic assumptions are the following:

(1) The resources R_i of queue i are uniformly distributed to all agents in the queue. As an immediate consequence, agent j receives R_i/L_i resources per time step when the queue length is L_i. The *generation time* g_{ij}, for agent j in queue i, assuming the queue length remains constant, is the length of time required for all components of the vector $r_jL_i/R_i = r_jL_i/R_i$ to exceed 1. Informally, the generation time is determined by the amount of time it takes the agent to acquire enough of each resource to make a copy of itself.

(2) The number of replicants a_{ij} of agent type j added to queue i each time step is equal to the number of agents j in the queue, L_{ij}, divided by j's average rate of resource collection: $a_{ij} = L_{ij}[R_i/r_jL_i]$.

(3) An agent j has a fixed probability d_j of dying at any time step, independent of how many time steps j has existed, so that the expected number b_{ij} of agent type j removed from the queue is $b_{ij} = d_jL_{ij}$.

The queued models examined here concentrate on what happens to the queues' lengths over time and on what kind of equilibrium, if any, arises. Real niches don't usually settle down in this way; however, there are useful intuitions to be gained from these simpler cases, and the simplifications have the advantage of making rigorous calculations possible. The calculations are built around the change in the number of agents of each type in the queues. Specifically, let D_{ij} be the *change* in the number of agents of type j in queue i on a selected time step. That is, D_{ij} equals the number of copies of j added, $L_{ij}[R_i/r_jL_i]$, minus the number deleted, d_jL_{ij}:

$$D_{ij} = a_{ij} - b_{ij} = L_{ij}[R_i/r_jL_i] - d_j L_{ij}.$$

When for all j in queue i all the D_{ij} equal 0 from one time step to the next, the queue length of i remains unchanged. In other words, the queue is in a steady-state equilibrium. The calculations for the following examples all proceed along these lines.

One queue with one agent type (crowding)

In this first model, all the agents are of the same type, so j = 1, and there is only one queue, so $L_{11} = L_1$. Under assumption 2 above, $a_{11} = L_1[R_1/r_1L_1] = R_1/r_1$ agents are added to the queue by replication. Under assumption 3, $b_{11} = d_1L_1$ agents are deleted. At equilibrium

$$D_{11} = a_{11} - b_{11} = 0,$$

so

$$R_1/r_1 - d_1L_1 = 0$$

and the equilibrium queue length is

$$L_1 = R_1/\ d_1r_1.$$

In niche terms, L_1 is the *carrying capacity* of the queue (the number of agents that can be supported by the resources provided). To give a numerical example, let the resource input R_1 be 100, the replication requirement r_1 be 4, and the "death rate" d_1 be 1/4; then $L_1 = 100$.

One queue with two agent types competing for one resource (competitive exclusion)

In this model, the total length of the queue L_1 is the sum of the numbers of the two agent types, $L_1 = L_{11} + L_{12}$. For the queue to remain the same length, it must be the case that

$$D(L_1) = D(L_{11}) + D(L_{12}) = 0.$$

One way to meet this requirement is to set

$$D(L_{11}) = L_{11}R_1/r_1L_1 - d_1L_{11} = 0,$$

where

$$L_1 = R_1/d_1r_1,$$

and, simultaneously, to set

$$D(L_{12}) = L_{12}R_1/r_2L_1 - d_2L_{12} = 0,$$

where

$L_1 = R_1/d_2r_2$.

For these simultaneous equations to have a common solution, either d_1r_1 must be equal to d_2r_2 (usually an uninteresting case because both agents are essentially identical competitors) or else one of the agents must disappear, yielding either

$\{L_1 = R_1/d_1r_1, L_{12} = 0\}$

or

$\{L_1 = R_1/d_2r_2, L_{11} = 0\}$.

In effect, there is a race in which one agent eventually replaces the other, an example of the phenomenon called *competitive exclusion*. If $D(L_{11}) > D(L_{12})$, the expected outcome is a queue consisting only of agent 1; if $D(L_{12}) > D(L_{11})$, agent 2 can be expected to prevail.

For a numerical example, use resource input rate $R_1 = 100$, and for agent 1 use the same values as in the first model: $r_1 = 4$, $d_1 = 1/4$. For agent 2 use $r_2 = 2$ and $d_2 = 1/4$. The result if agent 1 prevails is the same as in the first model: $L_1 = L_{11} = R_1/d_1r_1 = 100$. If agent 2 prevails, the queue length $L_1 = L_{12} = R_1/d_2r_2 = 200$. Note that the carrying capacity of the niche for agent 2 is twice that for agent 1. Which agent survives depends on early random deaths—if agent 2 has an "unlucky streak," with no copies remaining to produce offspring, it may be displaced early by agent 1.

One queue with two agent types, one collecting resources from the other (multiplier effects)

In this case, the resources available to agent 2 come through agent 1, as in a predator-prey or a parasitic relationship. A simple model of this interaction treats the probability P_{12} of interaction between agents 1 and 2 as a random "collision" (as in billiard-ball mechanics). Accordingly, the probability of an interaction between agent 1 and agent 2 depends on their proportions in the queue, L_{11}/L_1 and L_{12}/L_1:

$$P_{12} = 2r_{12}[L_{11}/L_1][L_{12}/L_1].$$

Here r_{12} is the fraction of interactions between agent 1 and agent 2 resulting in an exchange of resources. (Some collisions do not result in an exchange as in reaction networks; see section 4.4.) The total number of collisions per time step, cL_1, is set as fraction c of the total number of agents L_1 in the queue. In this case, there is an additional source of death for agent 1 from predation, given by

$$P_{12}(cL_1) = 2r_{12}[L_{11}/L_1][L_{12}/L_1](cL_1).$$

Summing the additions from replication (using the same calculations as in the first model), with subtractions for the two sources of death (predation and random death), gives

$$D(L_{11}) = L_{11}R_1/r_1L_1 - 2r_{12}[L_{11}/L_1][L_{12}/L_1](cL_1) - d_1L_{11}.$$

Each death of agent 1 supplies resources to agent 2 in the amount $P_{12}(r_1/2)$. With cL_1 collisions, the total input of resources to the queue for agent 2 is $P_{12}(r_1/2)(cL_1)$. As in the first example, then, the number of replicants of agent 2 is $P_{12}(r_1/2)(cL_1)(1/r_2)$ and, as the predator, agent 2 suffers only random death d_2L_{12}, so that

$$D(L_{12}) = 2r_{12}[L_{11}/L_1][L_{12}/L_1]\ ](cL_1)(r_1/2)(1/r_2) - d_2L_{12}.$$

The corresponding steady-state equations are

$$L_{11}R_1/r_1L_1 - d_1L_{11} - 2r_{12}[L_{11}/L_1][L_{12}/L_1](cL_1) = 0 \qquad (1)$$

and

$$2r_{12}[L_{11}/L_1][L_{12}/L_1]\ ](cL_1)(r_1/2)(1/r_2) - d_2L_{12} = 0. \quad (2)$$

Simplifying equation 1 yields the following:

$$\begin{aligned} &R_1/r_1 - d_1L_1 - 2cr_{12}L_{12} = 0, \\ &L_1 = R_1/d_1r_1 - 2cr_{12}L_{12}/d_1, \\ &L_{11} + L_{12} = R_1/d_1r_1 - 2cr_{12}L_{12}/d_1, \\ &L_{11} = R_1/d_1r_1 - [1 + (2cr_{12}/d_1)]L_{12}. \end{aligned} \quad (3)$$

Simplifying equation 2 yields the following:

$$\begin{aligned} &[L_{11}/L_1] - (r_2/cr_{12}r_1)d_2 = 0, \\ &L_1 = L_{11} + L_{12} = (cr_{12}r_1/r_2d_2)L_{11}, \\ &L_{12} = [(cr_{12}r_1/r_2d_2) - 1]L_{11}. \end{aligned} \quad (4)$$

Using equation 4 to substitute for L_{12} in equation 3 gives an equation for L_{11} in terms of system parameters:

$$L_{11} = R_1/d_1r_1 - [1 + (2cr_{12}/d_1)][(cr_{12}r_1/r_2d_2) - 1]L_{11}.$$

Multiplying out and collecting terms yields

$$\begin{aligned} L_{11} = R_1/d_1r_1 - [&-1 + cr_{12}r_1/r_2d_2 - 2cr_{12}/d_1 \\ &+ (2cr_{12}/d_1)(cr_{12}r_1/r_2d_2)]L_{11}. \end{aligned}$$

Dividing through by $cr_{12}r_1$ and recollecting terms yields

$$L_{11} = R_1r_2d_2/[(cr_{12}r_1)(r_1d_1 + 2cr_{12}r_1 - 2r_2d_2)].$$

For a numerical example, use the parameters from the previous numerical examples ($d_1 = 1/4$, $d_2 = 1/4$, $R_1 = 100$, $r_1 = 4$, $r_2 = 2$) and add the parameters $r_{12} = 1/2$ and $c = 1/2$. Then

$$L_{11} = 50/[(1)(1 + 2 - 1)] = 25$$

and

$$L_{12} = [1/(1/2) - 1]L_{11} = 25.$$

If the death of agent type 1 (the prey) is primarily from agent type 2, so that d_1 is close to 0, then

$$L_{11} = 50/[(1)(2 - 1)] = 50$$

and

$$L_{12} = [1/(1/2) - 1]L_{11} = 25.$$

Thus the predator, agent 2, benefits from the resources collected by agent 1. If agent 2, in turn, had a predator, agent 3, the resources would support still another agent. Thus, there is a multiplier effect. (See section 1.3.)

If the prey population is primarily limited by carrying capacity (set by R_1) minus losses incurred from predation, then the prey's population dynamics follows the rule

$$D(L_{11}) = aR_1 - bL_{11}L_{12},$$

where a and b are constants. If the predator population depends almost entirely on availability of prey, then

$$D(L_{12}) = cL_{11}L_{12},$$

where c is a constant. These two equations are a simple version of the Lotka-Volterra equations (Christiansen and Feldman 1986). Because of the $L_{11}L_{12}$ term, replication of the predator (agent 2) population approaches 0 when the prey population (L_{11}) approaches 0. When the prey population gets large ($L_{11}>>1$), the predator population begins to replicate rapidly. Because the predator population can increase only after the prey population has increased, a "lag" is imposed. The typical result of this lagged interdependence is continual oscillation of both populations, with the predator population lagging the prey population, as in the classic lynx/hare oscillation exhibited by the Hudson Bay Company's trapping records (Cole 1954).

Two arms with two agent types, one migratory and one not migratory (niche invasion)

Now, as was discussed in section 9.1, let agent type 1 have a tag that allows it to go to either arm I or arm II, while only agent type 2 resides in the queue at arm II. For simplicity, assume that a fixed fraction m of agent 1 migrates on each time step. Thus, a total of mL_{11} of the population of agent 1 in queue I migrates to queue II on each time step, and similarly mL_{12} is the number of agent 1 moving from queue II to queue I.

The equilibrium equation for agent type 1 in queue I gives

$$L_{11}R_1/r_1L_1 + mL_{21} - (m + d_1)L_{11} = 0.$$

Because only agent type 1 resides in queue I, $L_1 = L_{11}$, the equilibrium equation reduces to

$$R_1/r_1 + mL_{21} - (m + d_1)L_1 = 0, \quad (1)$$

where

$$L_1 = (R_1/r_1 + mL_{21})/(m + d_1).$$

For agents 1 and 2 in queue II, the equilibrium equations are

$$L_{21}R_2/r_1L_2 + mL_{11} - (m + d_1)L_{21} = 0 \quad (2)$$

and

$$L_{22}R_2/r_2L_2 - d_2L_{22} = 0. \quad (3)$$

Simplifying equation 3 gives

$$L_2 = R_2/d_2r_2. \quad (4)$$

Using $L_2 = L_{21} + L_{22}$ gives

$$L_{22} = R_2/d_2r_2 - L_{21}. \quad (5)$$

Substituting equations 4 and 5 in equation 2 gives

$$L_{21}d_2r_2/r_1 + mL_1 - (m + d_1)L_{21} = 0, \quad (6)$$

where

$L_{21} = mL_1/[m + d_1 - (d_2r_2/r_1)]$.

Multiplying both sides of equation 1 by $(m + d_1)$ and using equation 6 for the value of L_{21} gives

$(m + d_1)L_1 = R_1/r_1 + mL_{21} = R_1/r_1 + m^2L_1/[m + d_1 - (d_2r_2/r_1)]$,

$L_1[(m + d_1) - m^2/[m + d_1 - (d_2r_2/r_1)]] = R_1/r_1$,

$L_1 = (R_1/r_1)/[(m + d_1) - m^2/[m + d_1 - (d_2r_2/r_1)]]$.

To obtain a numerical comparison with the previous two-queue/two-agent-type model, use the parameters for the previous model, $d_1 = d_2 = 1/4$, $R_1 = 100$, $R_2 = 200$, $r_1 = 4$, $r_2 = 2$, and set the migration parameter to $m = 1/4$. Queue I contains only agent 1, so

$L_1 = L_{11} = 25/[1/2 - (1/16)/(1/2 - 1/8)] = 75$.

The population of agent 1 in queue II is

$L_{21} = 75/4[1/2 - 1/8] = 75(2/3) = 50$.

Thus, the ability of agent 1 to migrate increases its overall presence. If agent 1 is restricted to a single queue with no competitors, it will have $R_1/d_1r_1 = 100$ instances (one agent type/one queue). Under migration its total presence is $L_1 + L_{21} = 125$. In addition, the ability to move to queue II increases the resilience of agent 1, allowing survival if queue I "shuts down."

In another interesting setting of the parameters, agent type 1 requires *less* payoff than 2 to reproduce. Using all the same parameters, except that now $r_1 = 2$ and $r_2 = 4$, we get back to the second example, in which agent 1 displaces agent 2 in queue II, a typical example of an invasive species.

Summarizing the sequence

These models use simple mechanisms—gambling strategies and queues—to produce well-known, niche-related emergent phenomena—carrying capacity, crowding, competitive exclusion, resilience resulting from migration, and "multiplier effects" resulting from reuse of resources. By adding arms to the bandit, it is easy to extend this list to competition between generalists and specialists, mutualism and symbiosis, arms races, evolution of diversity in interaction networks, and even mimicry. However, there is more to be gained by first placing these models and their mechanisms in a broader framework using tags to determine interactions.

9.4 Recombination of tags within queues

Consider the bandits' arms as counterparts of regions in Dobzhansky's (1977) description of *biospace*: "[The] environment [is] divided into regions within which certain characteristic modes of adaptation are required . . . —the land and the sea, the forest and prairie, and the rocky shore and sandy beach are some obvious examples." This analogy can be moved into the framework of complex adaptive systems by following the suggestion at the end of section 9.1: assign tag-based conditions to the bandits' arms, much as was done earlier for tagged urns. For example, assign the condition ####00## to arm I, where the tag 00 starting at locus 5 designates a resource that can be tapped by an agent carrying that tag. The tag is a mechanism similar to a *ligand*—an active site that enables a protein to pass through a cell membrane (Alberts et al. 2007). In another sense, an arm with an assigned condition acts like a region of biospace that can be exploited by agents with appropriate tags.

Agent tag must match urn tag to enter queue.

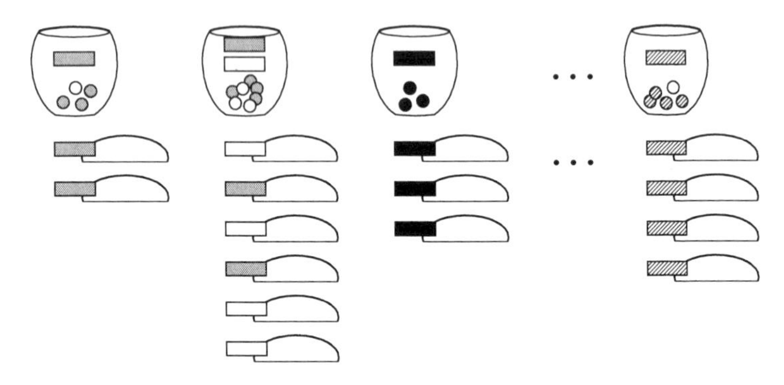

Figure 9.3
Urn niches.

It is easier to see the effects of tag modification with the help of a new notation for conditions and signals that emphasizes tags. In this notation, a single tag is specified by the locus at which it begins followed by the bits that specify the tag. For example, [(5)00] represents the tag ####00## . . . #. To represent two tags on a single string, two loci are specified, each followed by the specification of the corresponding tag. For example, [(2)1; (5)00] treats the string #1##00## as carrying two tags; of course, this arrangement could also be looked upon as the single tag [(2)1###00]. By bringing tags to the fore, this notation makes it easier to examine the effects of mechanisms like crossover and mutation on tag-mediated interactions within a niche.

Crossover between loci x and x + 1 is indicated by <x|>. So, for example, a cross at <4|> between a string carrying tag [(2)1] and another string carrying tag [(5)00], results in a pair of offspring, [(2)1; (5)00] and [-] (where [-] represents a string that carries none of the designated tags). If each tag provides entry to a different queue then the offspring [(2)1; (5)00] can exploit both queues—in comparison with its parents, it is a generalist. The other offspring, [-], is excluded from all three arms because it doesn't have any of the requisite tags. By recombining tags, then, offspring gain possibilities for exploiting new niches. Some offspring can exploit a wider range of resources (more arms); others may specialize to exploit interactions within a specific arm (niche).

When a particular arm requires a complicated tag for occupation it is more difficult to discover the tag that exploits that niche, but a "specialist" that does acquire that tag will typically have fewer migrants invading its niche. The specialist pays for this exclusivity by requiring more resources to replicate its more detailed tag (in the manner of Echo models). On the other hand, an arm that has a simple tag can be occupied by agents of many different kinds, and they use fewer resources to replicate their tag. The cost, of course, is that those agents will have many competitors, so they will receive a smaller share of the resources offered by the arm. Here, then, is a reflection of a classical distinction in population genetics: r-selection (outcompete rivals by exploiting readily available resources) vs. K-selection (outcompete rivals by specializing in the efficient use of limited resources)—see Christiansen and Feldman 1986.

Of course, some recombinant offspring will have tags not suited to in any niche—a net loss in the parent's fitness (reproduction rate). Still, in environments where the regularities uncovered by tags play a significant role, the overall rate of improvement under recombination is much higher than that obtained by uniform random variation (mutation) of the string's bits (as was detailed by the *schema theorem* in chapter 6). From a sampling point of view, the new tags offer plausible attempts at innovation because they are based on tested building blocks. Tags that do exploit new niches will spread rapidly under Darwinian selection, becoming parents for still further innovations. As Dobzhansky (1977) relates, "most families appear to have originated by a continuation of the same processes that result in the formation of genera. . . . For any given gene battery, effects . . . at a sensor site could range from deactivation . . . to an entirely new activating substance."

The two kinds of distributed control discussed at the end of section 9.1—migration based on resource availability and replication based resource collection—are easily implemented. In this model, agents migrate to new arms with a frequency that increases with decreasing resource collection in their current queue—"starving" agents are ready to move. Of course, they can migrate only to arms that accept their tag, but agents that can migrate to an under-occupied queue benefit from an increased reproduction rate. By examining the migration of tags, we can examine the competition between generalist agents and specialist agents in different kinds of niches.

Tagged arms are obviously closely related to the tagged-urn model, and, as we'll see in chapter 15, they can be analyzed similarly.

9.5 From bandits to arches

The object now is to examine these mechanisms in a more general signal/boundary framework. Two simple extensions increase the relevance of queued-bandit models to niches:

Instead of having the arms disperse a universal good ("cash"), let the each arm distribute a vector of the letters used to define tags and conditions. That is, each arm is treated as a site in the Echo sense, with a "fountain" of exogenous resources (letters) that can be used in constructing agents.

Instead of sharing resources proportionately among agents in a queue, use coupled reactions to distribute the resources. Then agents within a queue can exhibit the complex effects of recycling.

With these provisions there can be a greater diversity of agents within a queue, and queues can become longer. Agents can form coalitions (such as the caterpillar-ant-fly triangle used as an illustration in the Echo models), and there can be a variety of multiplier effects. Moreover, these buyer-seller sequences make possible credit-assignment routines, such as the bucket-brigade algorithm used in classifier systems (Lanzi 2000).

With these extensions, the queued-bandit models help stitch together the previous discussions of reaction networks, tagged urns, and recombination of building blocks. The models can be explored computationally, and, because they are tiered, they can be examined in stages of increasing complexity, offering a kind of "transparency" not easily attained otherwise. With recombination of tags, it becomes possible to study the evolution of interactions while avoiding simplifications that poorly characterize the circulation of resources in real signal/boundary systems (e.g., randomly generated networks or networks without feedback).

Though the stitching is relevant and suggestive, it still leaves us searching for an overarching framework. Unfortunately, even simple queued-bandit models are well beyond the broad analytic approaches offered by game theory or mathematical economics, so the path to generalization is unlikely to lie in that direction. An alternative path centers on using the tags in a queued-bandit model as the basis of a "signal/boundary grammar." We have seen repeatedly that tags serve a syntax-like function in directing signals and forming boundaries. Is there a grammar related to a finitely generated dynamics that will organize this lexicon? The next chapter explores that possibility by taking a closer look at grammars in their home territory, language.

10 Language: Grammars and Niches

10.1 Background

Chapter 9 suggested that grammars could extend the integration offered by queued bandits. Grammars and computer programs are closely related. Both grammars and programs produce solutions to complex problems by executing a sequence of elementary operations. As far back as the 1830s, Charles Babbage designed a programmable computer, the Difference Engine, by combining mechanical devices for addition and multiplication (Babbage 2010). Today's ubiquitous computers are a direct embodiment of Babbage's ideas. Similarly, grammars use simple steps to define complex objects, allowing both algorithmic and theoretical study of the objects defined. As a precursor to relating grammars and finitely generated dynamics, and as a way of developing some intuition about the possibilities of grammars, this chapter examines grammars in their "home territory," the study of language.

Grammars were initially the province of descriptive linguistics, familiar to most of us from high school exercises in the diagramming of sentences. In 1951, the logician Stephen Kleene introduced grammars to computer science by inventing a formal grammar to explain Warren McCulloch and Walter Pitts' groundbreaking but difficult approach to designing "logical circuits" (Kleene 1956). Grammars have had a central role in both theoretical and practical computer science ever since. In the late 1950s, Noam Chomsky brought grammars full circle by conjecturing that all natural languages are variations on a single universal grammar (UG), providing a formal science of linguistics in the process (Chomsky 1965). Under Chomsky's conjecture, each observed human language is obtained from a single universal grammar by setting (or learning) values for a set of variables, called *parameters*. Once the values for the parameters are set, all the rules of the universal grammar become specific, designating a particular language.

UG, with parameters set, describes a standard procedure for assembling elements (words) into *admissible* strings (sentences). However, UG does not tackle the multi-agent *cas* aspect of language. Because the multi-agent aspect of language is central to the study of signal/boundary systems, including language itself, this aspect of language is worth a closer look:

- Language is, above all, a social phenomenon, facilitating the interaction of agents.
- Language acquisition involves much more than determining parameter values through sampling. Babies and young children acquire language by using quite complicated procedures—gestures, shared attention, and the like.
- Each individual produces language in a distinct, idiosyncratic way. There are diverse individual grammars within the same language (Ke 2006).

The grammaticality of a string of utterances can be learned, in the absence of any training specific to grammars, by forming anticipatory (predictive) models. That is, strings of utterances that are effective in producing interactions will be used in similar situations, while strings that are ineffective will be discarded. Learning through anticipation plays an important role in the development of signal/boundary grammars (Holland, Tao, and Wang 2005).

10.2 Linguistic grammars

Language, fully utilized, is a capacity unique to humans among organisms on earth. It offers an enormous ability to signal a diverse array of environmental situations, both present and absent. We are still a long way from a comprehensive understanding of language, but grammars offer substantial help. In essence, a grammar helps to explain language's ability to produce sophisticated descriptions and communications while using a limited vocabulary. Contrast the use of grammatically sequenced utterances to using a large variety of single utterances to describe the same situations. Even if utterance repetitions are used, say, as a way of indicating urgency, the size of the vocabulary increases dramatically with any increase in the variety of salient environmental factors—size, form, color, direction, closeness, and so on. The ability to string utterances together in a meaningful way greatly reduces the requirements on vocabulary size.

As an example of the reduction in ambiguity offered by a grammatical sequence, consider a situation in which a red ball, a blue ball, and a cookie are placed on a small table. Suppose

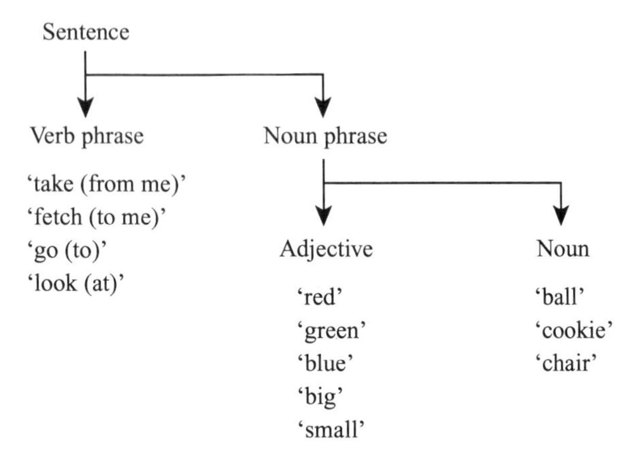

Sixty meaningful triples can be constructed from these utterances.

The number of meaningful n-tuples increases factorially as the number of utterances in each category increases.

E.g., increasing the number of utterances above to 20+20+20 yields 8,000 triples.

Figure 10.1
A grammar.

that a child (called L, for learner) is interacting with a teacher (called T). L may say "fetch," but T is left to decide whether it is one of the balls, the cookie, or even the table that is to be fetched. If L says "fetch cookie" there is no longer any ambiguity; however, if L says "fetch ball" there is still some ambiguity, which is resolved by the sequence "fetch red ball." If a separate single utterance were required for each case, the vocabulary would increase greatly. Figure 10.1 illustrates the point. With a simple Verb/Adjective/Noun grammar, a vocabulary of twelve utterances allows sixty meaningful distinctions. More generally, if there are n_1 utterances for the first position, n_2 for the second, and n_3 for the third, then the size of the vocabulary is $n_1 + n_2 + n_3$, but the number of triples (meaningful distinctions) increases factorially as $n_1 \times n_2 \times n_3$. For example, if $n_1 = n_2 = n_3 = 20$, there are 8,000 meaningful distinctions that can be made with this vocabulary of sixty utterances.

Observations of language acquisition

Without any explicit instruction in grammar, an infant progresses from babbling to a grammar-based language in a couple of years. What mechanisms make this possible? It is well established that the typical infant has "wired-in" capabilities involving vision, sound, and gesture. A newborn, without practice, imitates many of its mother's facial actions, such a sticking out the tongue. After a few months, an infant will direct its attention to any distinct object stared at by its mother. Infants strive for repeatability. Hand motion progresses from random waving to movement in a consistent direction across the visual field and then to purposeful touching. Babbling progresses from random sounds to simple repeatable utterances to phrases. And so on. The infant shows obvious joy at each successive advance in control. These capabilities are present in many pre-primate species, but human social interaction and learning parlay them into combinations that exploit experience to guide future action (anticipation). More and more evidence shows that the planned sequences of utterances characterizing language emerge from this combination (Bybee 2006; Five Graces 2009; Gao 2001).

Well-controlled experiments establish a toddler's progression from unreflective, present-oriented activities to assignment of labels to perceptual experience, and then to anticipation and short-term planning. In a typical experiment, a child is shown two target cards (e.g., a blue rabbit and a red car) and asked to sort a series of these cards according to one dimension (e.g., color). Then, after sorting several cards, the child is told to stop playing the first game and switch to another (e.g., shape: "Put the rabbits here; put the cars there"). No matter what dimension is presented first, three-year-olds typically continue to sort by that dimension despite being told the new rules on every trial (Zelazo, Gao, and Todd 2007). In contrast, four-year-olds recognize immediately that there are two sets of rules for the game and that a switch of rules is needed. Along similar lines, successive stages of linguistic development are closely related to increasing autonomy and to progress in control of utterances.

Consciousness, in its common-sense meaning, clearly expands as a child gets increasingly adept at using language (Dennett 1992; Hofstadter 1999). But "consciousness," like "life" or "mind," is difficult to define precisely. Nevertheless, there are well-developed sciences that center on difficult-to-define concepts such as "life" and "mind" (biology and psychology respectively), so we should not be too quick to dismiss an approach to language acquisition centering on consciousness. Using language as a tool for reporting conscious experiences can widen our view of the acquisition process. Zelazo, Gao, and Todd (2007) offer an interesting approach, postulating a hierarchical arrangement of consciousness based on age-dependent capabilities. Zelazo's observations can be modeled by means of a layered set of rules analogous to the approach used by Valentino Braitenberg in his 1984 book *Vehicles: Experiments in Synthetic Psychology*.

"Levels of consciousness" (LoC) models are based on metarules that are not language specific and are demonstrably available to pre-primates. Such models look for and combine mechanisms that have been either directly observed or conjectured on the basis of evidence. The models use common mechanisms much as engineers use gears and springs to understand the working principles of everything from watches to wagons, or as physicists speak of a "clockwork universe" when discussing causality. Though it is unlikely that simple laws will encompass the whole of language acquisition, even a rudimentary mechanism-based model should suggest testable possibilities for second-language teaching and for automatic language translation.

In this version of the LoC approach, each level of consciousness is obtained by adding a new mechanism to the previous level (Gao and Holland 2008). As earlier, rules are used to define two agents, L (learner—a newborn) and T (a language-competent teacher—say, a mother).

Level 0: unconscious activities (inherited cognitive abilities)

Pre-primate precursors to language acquisition

(i) Ability to imitate utterances and gestures.

(ii) Ability to distinguish between objects and actions.

(iii) Awareness of a mutually apprehended salient object or action.

(iv) Basic learning procedures, akin to Hebb's (1949) learning rule.

Typical rule

IF (T utterance) **THEN** (<imitate utterance>)

(Note that L will use its limited current abilities to attempt the match. For example, T-utterance "Gloria" can become L-utterance "Do-ee.")

Level 1: Minimal consciousness (innate reinforcement of repeatable activities, ranging from repetition of sounds and motions to actions that produce innate rewards)

Example: Directed motion of hand across visual field (a precursor to gesture)

Typical rule

IF (hand in vision cone) **THEN** (<move hand right>)

Level 2: Stimulus-response (conditioned) consciousness (labeling from long-term memory)

Example: Utterances that cause innate rewards (such as causing T to smile)

Typical rule

IF (milk bottle present) **THEN** (<utterance "milk">).

There will often be correlations between recurring patterns in the environment (e.g., correlations between actions and objects) that can be exploited through conditioning.

Level 3: Simple recursive consciousness (use of utterances to cause others to act)

Example: Utterances that lead to food acquisition when food visible

Typical rule set

IF (milk bottle present) **THEN** (<utterance "milk">)

T fetches milk bottle.

IF (milk bottle at mouth) **THEN** (<drink milk>).

Level 4: Extended recursive consciousness (use of labels to cause others to act when object is not present)

Example: Food acquisition when food not visible

Typical rule set

IF (hungry & no food visible) **THEN** <"milk">

T fetches milk bottle.

IF (milk bottle at mouth) **THEN** (<drink milk>).

Level 5: Self-consciousness (internally planned sequences of action, including sequenced utterances, that make it possible to look ahead and explore alternative courses of action)

Example: Distinguish between two similar objects using a sequenced pair of utterances

Typical rule set

IF (red ball present and blue ball present) **THEN** (internal signal x)

[The signal x is initiated by anticipated ambiguity.]

IF (internal signal x & red ball desired) **THEN** (internal signal y & <"red">)

IF (internal signal x & internal signal y) **THEN** (<"ball">)

(Note that this set of rules only allows the object word "ball" to be uttered *after* the modifier "red"—a simple form of proto-grammar.)

The fifth-level rules suggest how relatively simple rule sets can implement a grammar.

The "levels of consciousness" approach of Zelazo, Gao, and Todd (2007) has suggested new experiments and even new approaches to the teaching of language. Their approach is closely related to the manifesto produced by the Five Graces Group (2009). The Five Graces Group's manifesto centers on the idea "that patterns of use strongly affect how language is acquired, used, and changes over time." The effects of perceptual constraints and social motivation on the infant's increasing autonomy correspond to Zelazo's progression.

Models

As was emphasized earlier, the models being examined here don't try to parameterize the data; they are exploratory. Nevertheless, they are constrained by available data. They suggest how the data might arise, and they suggest further experiments that will clarify the data and the models. Because the models concentrate on social interactions (those between T and L and those of whole communities), they fit well within the framework of complex adaptive systems. As with other *cas*, the agents rarely settle down to a common static equilibrium in which all agents share a fixed grammar. Rather, the *cas* approach suggests that each agent will develop its own idiolect, though agents who interact regularly will have many common constructions in their idiolects. This, indeed, is a finding of modern linguistics (Ke and Holland 2006).

As with other complex adaptive systems, the linguistic signaling between agents involves both overt signals to other agents, and internal signals that coordinate utterances. The internal signals, as in classifier systems, need have no intrinsic meaning; they serve much like the uninterpreted bit strings that coordinate instructions in a computer program. The overt signals are such things as gestures and utterances that can be observed by other agents, as with the output of the rule IF (T lifts milk bottle) THEN (say, "milk"). The rule-based *cas* approach used here emphasizes the generation of sequenced utterances, or gestures, that (sometimes) steer other agents. Linguists call this aspect of language *production*. Generally the utterances are in response to a salient environmental situation. The use of grammar for production contrasts with its use to parse (understand) sequences produced by other agents, called *competence* by linguists. The difference between competence and production is familiar to anyone who can understand simple phrases in a language but cannot speak the language. Even sub-primate animals, dogs for instance, can be competent in understanding simple linguistic phrases, though they lack the ability to produce the phrases.

For a rule-based approach to be realistic, an agent must have many simultaneously active rules that emit internal coordinating signals. This simultaneous activity is roughly the counterpart of the simultaneous firing of assemblies of neurons in the central nervous system (Hebb 1949). Such coordination is critical to grammar formation, as in the level-5 rule systems mentioned above. As we'll soon see, changes in the tags of internal coordinating signals facilitate language learning.

In the LoC approach, the overall activity of the agent is conditioned on three factors:

A desire or need. An internal *need* signal is generated, for example, by a low reservoir. (See chapter 6.)

The current environmental situation. The agent's *detectors* generate a *environmental signal* string in which each locus indicates some property of the environment—for example, a salient object or an utterance.

An anticipation. The agent generates an internal *anticipation* signal string, indicated by a special tag, that mimics the environmental signal that is expected when the need signal ceases.

An utterance issued at any given point in time is typically conditioned upon signals of all three types. The agent's objective, in each situation, is to use a sequence of utterances to cause a desired change in its environment.

When an anticipated outcome doesn't happen, search and learning are activated. For example, suppose that L wants to hold a red ball that sits among other balls across the room,. L utters "ball" in the anticipation that T will fetch the ball. However, T fetches the wrong ball, so L begins some exploratory utterance sets. After several trials, L says "red ball" and T does indeed fetch the red ball. There are two deep questions here: What kind of learning procedure will enable L to make plausible trials? How does L reward the production of a successful utterance sequence? Mechanisms for addressing these questions, which will be discussed later, ultimately depend on the specification of the agent's system for controlling the sequenced utterances (its performance system).

Summary

The LoC examples above illustrate the way various combinations of two utterances, three utterances, and so on, can provide substantial refinements in expression and meaning. At the child's single-utterance stage, the utterance "milk" can have various implied meanings: "Give me some milk," "Look at the milk bottle over there," "My sister has the milk bottle," and so on. Combining the two utterances "give" and "milk" greatly reduces this ambiguity. In mathematical terms, we refine a broad equivalence class into a set of smaller, more informative subclasses. Language acquisition, looked at in this way, is a system that adds generative rules as the learning agent's experience refines its levels of consciousness. The result is an acquired set of generators (vocabulary) and a progressively refined set of generating rules (a grammar). The next chapter will extend this approach to other signal/boundary systems.

10.3 An example

Here is a simple example of an agent that operates according to LoC precepts. The agent is organized as depicted in figures 10.2 and 10.3.

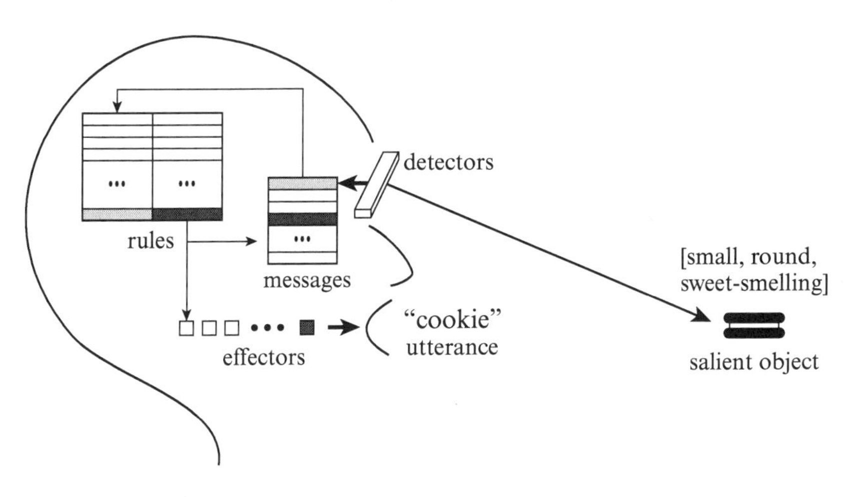

Figure 10.2
A language-using agent.

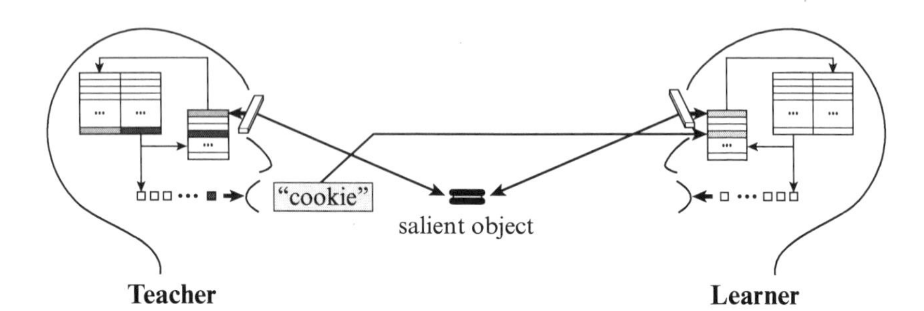

Roughly: Teacher has the rule IF [cookie observed] THEN [say "cookie"].

More carefully:

The detectors produce a message m1 tht records the properties of the observed object (small, round, sweet-smelling, ...).

The rule has a condition (IF) part that is satisfied by m1, and an action (THEN) part that sends message m2.

The message m2 activates the effectors that produce the utterance "cookie".

Figure 10.3
Interactive agents.

The positions in a signal string have the following interpretations:

Detectors

Nine loci in the detector string are allocated to properties of a salient environmental pattern. Three additional loci are used to designate different desires, with the following intended interpretations:

d_1	d_2	d_3	d_4	d_5	d_6
mother	cookie	red	green	blue	big
d_7	d_8	d_9	d_{10}	d_{11}	d_{12}
small	ball	chair	desire to grasp	desire to eat	desire to sleep

Internal signals

These strings use twelve loci to serve "grammatical" purpose (recalling that these loci only have "meaning" in terms of the rules they activate). Tags distinguish between vocabulary categories (see figure 10.1):

{v-'fetch', v-'take', v-'go', v-'look'},

{a-'red', a-'green', a-'blue', a-'large', a-'small'},

{n-'ball', n-'cookie', n-'chair'}.

The loci in an internal signal string have the following intended interpretation:

s1	s2	s3	s4	s5	s6
v-'fetch	v-'take'	v-'go'	v-'look'	a-'red'	a-'green'
s7	s8	s9	s10	s11	s12
a-'blue'	a-'large'	a-'small'	n-'ball'	n-'cookie'	n-'chair'

Effectors

The agent has twelve distinct effectors, one for each utterance (i.e., there is no parsing of the utterances into phonemes). <fetch> indicates that the effector, when activated by a signal, produces the utterance "fetch," and so on.

e1	e2	e3	e4	e5	e6
<fetch>	<take>	<go>	<look>	<red>	<green>
e7	e8	e9	e10	e_{11}	e_{12}
<blue>	<large>	<small>	<ball>	<cookie>	<chair>

Tags

A prefix of five bits is added to each string to serve as a tag specifying the origin of the remaining twelve bits in the signal:

y1	y2	y3	y4		
detector	internal	internal	internal		
y5					
effector	signal	v-signal	a-signal	n-signal	signal

Signals

With the interpretations listed above, signal strings of length 17 are adequate to provide all the interpretations listed. The first five tag bits determine the interpretation for the remaining twelve bits. As a specific example, use 10000 for tag y_1, 01000 for tag y_2, and so on; then the tag is 10000 indicates a string with the next twelve bits interpreted as a detector signal $d_1d_2d_3 \ldots d_{12}$.

At time t, let D(t) be the detector signal, let I(t) be an internal coordinating signal, and let E(t) be an effector-activating signal. Then a typical rule for this agent has the form

IF D(t) & I(t) **THEN** E(t).

This rule can be compactly described using classifier notation. To make specification of grammars by means of these rules easy, add the *pass-through* symbol introduced for reaction networks, so that selected symbols in the rule's incoming signal are passed through to the outgoing signal. In the present context, wherever a 'don't care' symbol # occurs at the same locus in *both* the condition and the action parts of the rule, then whatever symbol appears at that locus in the incoming signal is passed through to the outgoing signal. For example, the signal **111000** is accepted by the rule

IF ##1#0# **THEN** ##0#1#

and the outgoing signal is **110010**.

To continue the example, let h be an internal signal sent when the agent is hungry; let d(c.rb,bb) be a signal sent from the detectors that indicates a cookie, a red ball, and a blue ball sitting on a chair in the visual field; and let s be an internal signal that initiates “sentence” production. With the tags v (“verb”) and a (“adjective”), and the pass-through notation, the following set of rules implements a piece of grammar causing a ‘verb-adjective-noun’ utterance sequence:

IF h **THEN** s

IF d(c,rb,bb) & s **THEN** x & v-‘fetch’

  IF #‘fetch’ **THEN** <“fetch”>,

where

IF v# & x# **THEN** y & a-‘red’

  IF #‘red’ **THEN** <“red”>

IF a# & y# **THEN** n-‘ball’

  IF #‘ball’ **THEN** <“ball”>

In these rules, ‘fetch’ designates a part of an internal signal, while <“fetch”> designates an overt utterance produced by an effector. The grammatical tags determine the order of the utterances, as in LoC level 5 in section 10.2.

10.4 Linguistic networks—speech communities

A speech community consists of multiple agents that interact through mutually intelligible, sequenced utterances. Speech communities amount to dialectal niches formed within the human moiety. The niches so formed give rise to richer versions of the "pack" phenomena observed in earlier primates, and other animals, dividing the world into "them" and "us," with all the advantages and disadvantages that accrue from such simplifications.

Networks are a natural way to represent these dialectical niches. It is advantageous to embed such networks in the *cas* framework because that framework can handle a broad range of agent-based linguistic data: observations of language difference within agents and across speech communities, signal processing observed in agent-oriented psychological and neuroscience experiments, levels of consciousness transitions within an agent, and anthropological and historical studies of grammaticalization. Moreover, these observations, once described within the *cas* framework, become subject to computer-executable modeling. Computer modeling makes it possible to prove, at least in principle, that specific mechanisms can combine to produce the selected observations.

Language exists both in individuals (as idiolect) and as an emergent phenomenon in the community of users (as dialect). Linguistic diversity is so great, even within a given language community, that a single average or ideal agent cannot reasonably represent the community. Accordingly, any attempt to understand the language community must confront the fact that each agent has different cognitive abilities, including differences in "wired-in" abilities. This difficulty is compounded by the fact that there is little or no central control of language use, even where a national body attempts to define proper usage. As with markets, there is no ideal ("fully rational") participant. However, an agent-oriented *cas* model, with its distributed control, is well suited to studying this linguistic diversity.

Though language use is distributed and diverse, patterns are pervasive, emerging in a manner reminiscent of the way bird flocks and fish schools arise and persist without central control. Some of these persistent patterns serve to generate linguistic organization—phonemes, words, sentences, discourse, and so on. There are also dynamic patterns that are found within most agents in the community—stages in acquisition, pidgin formation, the appearance and disappearance of idioms, and so on. Some of these patterns arise from interactions that exploit a newborn's wired-in primitives. Once again, "items that are used together are fused together" (Bybee 2006). Later in language development, the amalgam of utterances, gestures, and situational manipulation is strongly constrained by an intensely interactive social existence. There seems little likelihood of understanding this continuing interplay in the absence of a *cas*-style description.

The language community becomes ever more complicated because linguistic signaling makes long-range anticipation possible. Plans can be laid and investigated before any execution takes place. This contrasts with most animal signaling. A hunting pack uses subtle signals involving many senses (vision, sound, touch) to facilitate joint action, and there can be substantial abstraction (e.g., there is little direct relation between a warning cry and the form of the danger). However, each signal/action combination immediately conditions the next round of signals, in an almost Pavlovian way (Braitenberg 1984). A hunting pack acts with localized, action-directed foresight. In contrast, language makes it possible to examine multi-step options ahead of action. The resulting social subtleties are legion, ranging from agriculture to war plans and social soirees.

What options are there for representing these subtle linguistic interactions? More important, what options are there for representing the continuing change of language in both individuals and communities? Though networks are useful tool for representing interactions, formative interactions, such as language acquisition, involve a temporal sequence of modifications. Thus, a representing network is an evolving entity, many of its changes depending on the tags (e.g., grammatical endings) that determine grammatical constructions. The changes involved are not easily defined by rules that apply only to network structure (such as adding new nodes and connections to nodes that already have a high number of connections).

10.5 Mechanisms of language formation and change

As with niche formation (chapter 9), then, mechanisms of formation and change play an important part in understanding language. Language acquisition depends on steady learning from the sequenced utterances and actions of members in the language community, with ever-increasing complexity in a learner's ability to produce meaningful utterance sequences. Language also changes over longer historical periods, such as the changes in English since the time of Shakespeare.

The LoC language models of the preceding sections examine learning implemented by changing the conditional rules employed by the learner. Here, it is useful to recall the two modes of learning in rule-based classifier systems: credit assignment (which favors rules with tested efficacy) and rule discovery (which generates new rules to be tested). The credit-assignment mode relies on sampling extant rules in various situations, weighting them according to the value of the outcome; the rule-discovery mode requires methods for generating new rules (new hypotheses), as was discussed in chapter 6.

In the LoC models, a small, fixed set of operators can generate quite sophisticated rules for language behavior. Indeed, the "wired-in" evolutionary inheritance from pre-primates provides a powerful foundation for generating new language rules, even though these capabilities are not language-specific. The random variation and imitation that accompany the lowest levels of consciousness provide a random sampling that helps uncover the most primitive building blocks (say, phoneme-like utterances). Sounds and gestures reinforced by a teacher T become the building blocks for more complex utterances and gestures. Bybee's (2006) remark that "items that are used together fuse together" offers a good example of an operator that creates building blocks for a new level. A grammar, then, can be understood as a network built up from the categorized instances of language use. The nodes of the network designate linguistic elements (e.g., words), and the connections designate ways of sequencing the elements into sentences.

Increasing complexity and consciousness come from recombining building blocks already confirmed. The schema theorem (chapter 6) points up the way in which building blocks for rules persist when the rules carrying them offer advantages (in the present case, within the language community). Building blocks, like grammars, offer combinatoric possibilities. The next chapter discusses in detail how a large variety of useful or meaningful structures can be constructed from a small number of well-chosen building blocks. Then, selected combinations of building blocks that work well together at one level become the building blocks for the next level. Moving up the LoC hierarchy to produce more complex behaviors is a much more efficient process than trying to "establish" complex rules, one by one, at the highest level of consciousness. Tags, as demonstrated by the examples in section 10.3, offer a direct way of exploiting the combinatoric possibilities. Tags are particularly relevant to the grammar-like generation of *sequences* leading to valued culminating activities (a mother's smile, a prized object brought within grasp, and the like). This scenario, as we'll see, fits well within the *cas* framework.

10.6 The consequences of change

When a grammar's rules (or their equivalent) are inferred, rather than being inherited, it might be expected that each agent will have its own idiosyncratic grammar. Observation confirms this conjecture (Ke 2006). Thus, we get a hierarchy of niches: idiolects, dialects, language communities, and language groups. Within this hierarchy, language provides long-range anticipation and increased agent autonomy, so that an agent's behavior is not conditioned on current stimulus alone. Instead, current stimulus only modulates ongoing internal activity (such as playing through possible scenarios mentally). That is, an agent can "run" an internal model that permits "lookahead" and an internal exploration of options. Internal models (their construction and structure) are discussed in detail in *Induction* (Holland, Holyoak, Nisbett, and Thagard 1986). Salient cues within an internal model strongly influence the overall interpretation of utterance sequences (Five Graces 2009). Thus, communication often consists of prefabricated sequences based on anticipations, rather than word-by-word choice.

Much of the theoretical work in computer science centers on the well-defined class of hard computational problems. Recently, within this class of hard problems, it has been documented that there is a sharp transition between "obviously solvable" and "clearly unsolvable" problems. That is, the "really hard" problems occupy a small portion of the whole set. This result bears on transitions in language ability (Five Graces 2009). An accumulation of multiple small phenotypic differences between humans and other primates—degree of sociability, ability to share attention, memory capacity, rapid sequencing ability, vocal tract control, and so on—could produce a sharp transition in ability to communicate. The well-studied formalism of finitely generated systems contains grammars as a special case. Finitely generated systems, which offer a way to study the profound consequences that can be induced by a combination of small changes, are the subject of the next chapter.

11 Grammars as Finitely Generated Systems

The transferences of force from agent to object, which constitute natural phenomena, occupy time. Therefore a reproduction of them in imagination requires the same temporal order. . . . Things are only the terminal points, or rather the meeting points, of actions, cross-sections cut through actions, snap-shots.

—Ernest Fenollosa

11.1 Introduction to finitely generated systems

Chapter 10 examined grammar's power to use a small vocabulary to define a large number of complex, descriptive sentences. The present chapter uses the mathematical formalism of *finitely generated systems,* to develop a rigorous, overarching description of such assemblies.

A finitely generated system consists of a set of generators,

$G = \{g_1, g_2, \ldots, g_n\},$

and a set of *operators* R for assembling the generators. The operators are applied repeatedly to partly assembled structures—strings of generators—to produce a corpus of structures (such as chromosomes). For a finitely generated system to be interesting, the number of distinct generators should be small relative to the number of structures generated, though any number of copies of the generators can be used in the process. The principal advantage of a finitely generated system, in both theory and modeling, is its ability to provide a compact, unifying definition of a large (perhaps infinite) corpus.

It is noteworthy that a finitely generated system can provide a hierarchy of structures (e.g., subassemblies), each with a different generated subsystem, also describable by a *finitely generated system*. Language provides a familiar example. Individual letters (written language) or sounds (spoken language) are sequenced to yield words; words become the generators for a corpus of sentences; sentences, in turn, provide paragraphs; and so on. Herbert Simon (1996) made good use of this possibility in his well-known watchmaker parable. Two watchmakers use two different methods to construct a watch. The first watchmaker constructs the watch by adding one piece at a time until the watch is completed; the other watchmaker constructs subassemblies and then combines the subassemblies into larger subassemblies. If there is some chance that an assembly will "fall apart," then the second watchmaker loses only the time put into a subassembly, while the first watchmaker may lose the time required to assemble most of the watch. Simon gives an insightful analysis of construction times and stabilities under the two methods of assembly. The *finitely generated system* framework helps to extend Simon's ideas to the broader signal/boundary context.

It is *not* the object of a finitely generated system to capture details of the *mechanisms* (operators) used to assemble generators. For example, if we look upon the elements of the periodic table as generators for the corpus of chemical compounds, the object of the periodic table of the elements is not to capture the details of quantum-mechanical bonding that determines the table's organization; valances suffice as the rules of connection (operators) for assembling the elements. If we look at Babbage's computer, it is enough to know that it provides for sequences of addition, multiplication, and other operations without having to know the particular mechanisms Babbage used to make that possible. In short, with a finitely generated system, we can study the interactions defined by the operators without knowing the mechanisms that underpin them.

A finitely generated system for signal/boundary systems, then, should provide a way to explore the generation of signal and boundary hierarchies that employ conditional interactions based on tags. The generators are the letters used to define tags in addition to the letters used to define the rules and signals that employ the tags. To be relevant to signal/boundary structures, the corpus so defined must provide for (i) extensive agent-like interaction, (ii) learning and adaptation through modification of structure, and (iii) evolving diversity. When used with appropriate operators, the alphabet used for classifier systems and tagged urns supplies relevant signal/boundary generators, as we'll see. However, it is helpful to look first at an easy-to-define, rigorous example of a finitely generated system: *finitely generated groups*. The next section, intended for those who want to add rigor to intuition, defines finitely generated groups but it is not obligatory for understanding the exposition in later sections.

11.2 A precise example: finitely generated groups

Though the generators and rules used to define finitely generated groups are concise and easy to use, it is another matter to understand the structures so generated. In this, finitely generated groups are similar to games such as chess and Go. Being able to use the rules is a long way from being able to understand the patterns and structures they generate; understanding the structural implications takes an expert's experience and insight. Still, without being an expert, one can gain a feeling for what the game or group is like by looking for recurring patterns—building blocks.

The starting point for *finitely generated groups* is the general mathematical concept of a *group*, a straightforward generalization of arithmetic multiplication. A group G consists of a set of elements $E = \{e_1, e_2, \ldots, e_j, \ldots\}$ (say, numbers), with an operation $\times$ (say, multiplication) that applies to all the elements in E. For G to be a group, the $\times$ and E must satisfy the following requirements:

(i) G contains an *identity* element e, such that for any e_i belonging to E

$e_i \times e = e \times e_i = e_i$ (for example, in arithmetic, $3 \times 1 = 1 \times 3 = 3$.)

(ii) For each e_i belonging to E, there is an *inverse* element e_i^{-1} such that

$e_i \times e_i^{-1} = e_i^{-1} \times e_i = e$ (for example, $3 \times (1/3) = (1/3) \times 3 = 1$).

(iii) For any e_i and e_j belonging to E, the product $e_i \times e_j$ also belongs to E. (For example, if 3 and 2 belong to E, then $3 \times 2 = 6$ also belongs to E.)

(iv) The operation $\times$ is associative, so that $e_i \times (e_j \times e_k) = (e_i \times e_j) \times e_k$.

Because the group has only the single associative operator $\times$, one can write products as strings without explicitly writing the operator, so that $e_i \times e_i^{-1}$ becomes $e_ie_i^{-1}$, $(e_j \times e_i) \times e_i^{-1}$ becomes

$e_j e_i e_i^{-1}$, and so on. Though these requirements are simple and transparent, most groups have intricate structures with unusual properties (think of the fact that any number can be represented as a product of prime numbers).

Finitely generated groups are particular kinds of groups defined by first selecting a finite set of elements $G = \{g_1, g_2, \ldots, g_k\}$ from E. G is typically a small set. G must contain the identity element e, and G must also contain the inverse g_i^{-1} of every element g_i belonging to G; thus, G has the form

$$\{e, g_1, g_2, \ldots, g_b, g_1^{-1}, g_2^{-1}, \ldots, g_b^{-1}\}.$$

G, in this form, is called a set of *generators* for the finitely generated group; strings (products) over G are elements of the group.

Now let G* be the set of *all strings* (products) that can be formed from the generators in G:

$$G^* = \{e, g_1, g_2, \ldots, g_b, g_1^{-1}, g_2^{-1}, \ldots, g_b^{-1}, g_1g_1, g_1g_2, \ldots, g_b^{-1}g_b^{-1}, g_1g_1g_1, \ldots\}.$$

G* is called a *free finitely generated group*. G* clearly consists of all possible *distinct* strings of generators, noting that $f_1f_1^{-1}$ is replaced by e wherever it occurs. For example, the string $f_1f_1^{-1}f_2$ is treated as identical to the string f_2 because $f_1f_1^{-1}f_2 = ef_2 = f_2$. Consider, then, the free group A* based on just three generators: $\{e, a, a^{-1}\}$. Because any string of a's, no matter how long, belongs to A*, the group consists of an infinite number of elements. Note, however, that the string $aaa^{-1}a^{-1}a$ is identical to the string a under the rules of the group. That is, many different strings can represent the same element of the group. This observation introduces a structural question: Is there a standard *canonical* form for uniquely representing a set of equivalent elements in A*? For the free group A* it is easy to show that equivalent elements can be reduced to one of the following forms: e, a string of the form aa . . . a, or a string of the form $a^{-1}a^{-1}a^{-1} \ldots a^{-1}$. (To arrive at the canonical form, simply carry out all cancellations in the string). Knowing that canonical forms exist makes it much easier to understand the effect of the product operation on the elements of A*. In other words, the canonical forms give insight into the structure of the group.

The definition of a *free* finitely generated group, can be extended to the general definition of a *finitely generated group*. To do this, add to the four rules defining G* a set of additional rules C, called *constraints*. Each constraint in C has a role similar to rule ii above, selecting a pair of elements (strings) in the group (say, y and z) that are to be equated, so that y = z. The requirement can be rewritten as $yz^{-1} = e$, and all constraints in C can be put in that form, so that the right-hand side of the constraining equation is always e. For example, let the A* be constrained by a set C_1 of two equations:

aaaa = e

and

$a^{-1}a^{-1}a^{-1}a^{-1} = e$.

Designate the result $A^*|C_1$. Then whenever a string x in A* contains a substring of the form aaaa, the substring can be replaced by e, which in turn has the effect of removing the substring aaaa from the generated corpus. For example, the string $a^{-1}a^{-1}aaaaa^{-1}$ is identified with the string $a^{-1}a^{-1}a^{-1}$. By doing all such cancellations, we go from the infinite number of elements in A* to a group $A^*|C_1$ with exactly four elements:

$\{e, a, a^{-1}, aa\}$.

These elements serve as a *canonical* presentation of $A^*|C_1$. The same technique can be used to determine a canonical form for any finitely generated group.

The canonical forms for $G^*|C$ can be quite complex, even when C is small. We have just seen that two equations are enough to take the infinite group A^* to the group $A^*|C_1$ with only seven distinct elements. To get a feeling for the difficulty, ask what the canonical form is for the elements of

$$G^*|C = \{e, a, a^{-1}, b, b^{-1}\}|\{aba^{-1}ba^{-1}b^{-1} = e\}.$$

General structural problems for finitely generated groups are so complicated that comprehensive structural descriptions have only recently been achieved (Birkoff and MacLane 2008).

Though it is a stretch, we can think of the canonical representative of a set of strings as the "meaning" of that set. Looking back to the grammatical presentation of language (chapter 10), we could think of a word as a kind of "canonical form" representing the class of situations to which it applies. The meaning of a grammatical string of words would then be the intersection of classes specified by its words.

11.3 Dynamic finitely generated systems

In a finitely generated system the operators are fixed. How, then, does such a system handle change and learning?

Generated programs

Here it is helpful to think of a simulation (say, a video game) from the viewpoint of finitely generated systems. Ultimately, the simulation is specified by a sequence of instructions in a computer. A conventional computer, at the circuitry level, consists of a numbered set of registers, each of which holds a binary string that can be interpreted either as a piece of data or as an instruction. When the string is interpreted as data, it typically serves as a number; when it is interpreted as an instruction, the first part of the string specifies an operator (e.g., addition), and the second part serves as the address of a register to which the operator is to be applied. The set of operators designed into the circuitry is small, typically 32 or 64 operators. A program, then, is specified by a sequence of registers that are to be interpreted as instructions. There are, of course, layers and layers of organization sitting atop this simple base, but in the end the operation of the computer, no matter how sophisticated the software, is determined by this base. From this "assembly" viewpoint, then, a program, and the dynamic it implies, is generated by a sequence of instructions based on the small set of pre-specified operators. The set of all sequences that can generated by stringing together allowable instructions constitutes the corpus of programs.

The important point about a computer program, for present purposes, is that it specifies a dynamic that is realized as the program is executed step by step. The dynamic can be quite complicated because *conditional operators* cause jumps from one part of the program to another—IF (the result of the previous step is x) THEN (jump to step y). Conditional operators make subroutines possible, and the execution sequence can repeatedly jump back to subroutines previously executed, so that a finite program can produce a continuing dynamic. Moreover, one part of the program can change the content of registers that contain another part of the program, so programs can modify themselves. A vast range of dynamics can be generated in this way. Ultimately, when the operators are carefully chosen, *universality* results. For any well-defined algorithm, there is a program in the corpus that executes that algorithm. Of particular interest for studies of signal/boundary dynamics, a universal corpus contains algorithms that learn or adapt. As a result, there are programs in the corpus that can simulate the coevolution of the signals and boundaries of adaptive agents.

Generated populations

The focus here is on the *step-by-step* execution of the generating procedure, rather than on the usual concentration on the members of the completed corpus (the set of well-defined programs, sentences, or the like). That is, successive changes in the corpus are placed at center stage. One way to provide this emphasis to treat subsets of the corpus as *populations* that change as the generation procedure is executed. The generation procedure starts from an *initial population* of strings X(0). X(0) is modified by selected operators from R to produce a new population X(1), a new *generation,* that contains some new strings. Typically, the new strings replace strings already in X(0), though that isn't a necessary requirement. An appropriately designed replacement process can implement a conservation of generators, thereby satisfying the conservation-of-elements requirement for the reactions (discussed in chapter 3).

In going from one generation to the next, the elements used by the operators can be selected at random or in a prescribed order. Each selection procedure yields a distinct dynamic. Once X(1) is formed, any within-string identities imposed by the constraining relations C (as described in the preceding section) are removed so that X(1) contains only canonical forms. The whole process is then iterated to produce X(2), X(3), and so on. The result is a dynamic finitely generated system (hereafter abbreviated to *dgs* for convenience). By treating coevolution in the *dgs* framework, we gain a uniform way of examining commonalities of coevolution across the full range of signal/boundary systems.

Progressive adaptation

When a *dgs* uses the alphabet {1,0,#} to define tags and conditions, as in the discussions of tagged urns, then successive generations provide a specific description of the system's adaptations. To implement an implicit fitness requirement (chapter 6), the *dgs* must contain operators that replicate strings in X(t) only when the required generators are locally available. That is, there must be operators in the *dgs* that reassemble an agent's collected strings into strings that replicate the strings defining the agent's structure. With this provision, successive generations X(t) under the *dgs* will contain agents that are increasingly adept at collecting and retaining the resources required for their replication.

Anticipation

There is an additional requirement. In language, in music, and even in visual saccades, each signal produces expectations for the next signal:

In language, IF (one of us utters a word) THEN (any preceding conversation produces expectations for the words to follow). These expectations are so strong that it is still possible to guess the meaning ni wtretin Esngilsh wneh teh oderr fo het leertts si sbcrlmabed.

In music, a given chord produces an anticipation of a harmony to follow. Even for a symphonic movement, the combination of a key and a "long line" produces an expectation of the way in which the music will unfold.

In mammalian vision, the eye jumps (saccades) from location to location, rather than using a uniform scan. Each "stop" produces a very local high-resolution "snapshot," accompanied by a blurry peripheral image. This combination provides an expectation (often unconscious) of what the next "snapshot" in the series will show. The expectation is easily confirmed by the "startle reaction" that occurs when the next snapshot is entirely unexpected.

These examples emphasize grammar's role in generating anticipations by determining what building blocks can be added to those already assembled. That anticipations depend on the succession(s) generated by the grammar again puts emphasis on the step-by-step execution of the *dgs*.

Internal models

Anticipation becomes much more complex when an agent can run an internal subroutine that simulates some part of the environment (perhaps a part that includes other agents). Using an *internal model* (Holland, Holyoak, Nisbett, and Thagard 1986), the agent can explore various sequences of action before committing to the first action in a chosen sequence. To make this virtual exploration possible, the program implementing the internal model must *not* react directly to environmental inputs—it must be autonomous during its exploratory phase. One way to accomplish this autonomy is to tag the internal model's signals so that the model's execution proceeds in parallel with other internal signal processing. Tags for the internal model's signals act similarly to the tags that enforce certain sequences in a program that defines a linguistic grammar (as was discussed in section 7.6). Internal models, so used, greatly enhance the agent's range of anticipations, contributing to its level of consciousness (as was discussed in chapter 10).

11.4 Using a *dgs* to study signal/boundary coevolution

The task now is to specify a *dgs* framework that meets the requirements set forth in section 7.6:

a set of generators and generating rules that define signals and boundaries by combining building blocks,

spatial distribution of signals and boundaries,

the wherewithal to produce arbitrary signal-processing programs,

Darwinian selection by limiting the reproduction of an agent to collection of (copies of) the generators that specify the agent's signal/boundary structure,

and

provision for agents to have internal models that can be run autonomously to provide anticipations.

The next three chapters will undertake this task.

12 An Overarching Signal/Boundary Framework

12.1 The advantages of unification

A good way to begin the discussion of unification is by looking at the history of algorithmic problem solving.

Algorithms, such as routines for determining the volume of a solid, have been known since early Egyptian times. However, as the problems and the related equations became more difficult, the use of algorithms became more laborious. Various special-purpose machines were invented to ease the labor, but not until the first part of the nineteenth century did the concept of a single machine capable of different kinds of calculation come to the fore. Babbage's *difference engine*, an ingenious combination of mechanical devices for addition and multiplication, could be *programmed* to execute algorithms for constructing a variety of function tables useful in solving equations. This idea of programmability, carefully developed by Ada Byron (Lady Lovelace), led to the concept of a "machine language," with a "grammar" that specified allowable ways of sequencing the operators to produce programs for the machine (Babbage 2010). The critical advance to a general-purpose computer came when

Alan Turing (1936) specified a set of operators (roughly, addition, negation, and IF/THEN operators) that ensured the computer's universality. The result was a unification of disparate, often ad hoc techniques that led to important advances in problem solving.

The objective of a *dgs* (dynamic generated system) framework for studying signal/boundary systems is similar. Find *dgs* generators and generating rules that open the way to modeling *arbitrary* signal/boundary systems, with particular emphasis on the coevolution of signals and boundaries. The general-purpose computer example makes it clear that such an attempt can be more than an academic exercise.

Consider first the advantages for signal/boundary theory of defining agents and interaction networks via generators. This mode of definition opens the way to studying the coevolution of signals and boundaries by studying the successive recombination and modification of tags treated as generators. By using a genetic algorithm on the tag strings, we can compare effects of recombining tags in different settings (as ligands, active sites, phenotypic markers, case markers, and so on), and can explore the emergence of motifs, signals, hierarchies, and the like, under evolutionary pressures. This formulation leads on to the study of the emergence of niches in different complex adaptive systems (ecological communities, dialects, industrial complexes, etc.). On a still larger scale, we can look for common origins of diversity and boundary hierarchies under the coevolution of signals and boundaries.

When commonalities are observed in these models, they suggest experimental tests for underlying mechanisms. To quote George Uhlenbeck, "general phenomena [liquid-vapor phase transition, triple point, etc.] . . . must have a general explanation: the precise details [of molecular structure and intermolecular forces] should not matter." When an overarching structure is mechanism-generated and makes the use of well-developed parts of mathematics possible, the gain is indeed substantial. This chapter examines the possibility of such a mechanism-generated *dgs* framework for signal/boundary systems.

It is worth emphasizing again that the object of exploratory *dgs* models is not to capture detailed mechanisms of boundary formation, any more than the object of elementary chemistry is to capture the detail of quantum-mechanical bonding. Rather, the object is to explore the possible interactions of basic mechanisms that are relevant to the study of signal/boundary systems. We want to follow a path like Babbage's, moving from the mechanical elements we have examined to a program-like, grammatical way of describing the evolution of tag-defined signal/boundary networks.

12.2 Configuration of a unified signal/boundary model

The overall goal of the *dgs* described here is to generate, via coevolution, boundary hierarchies corresponding to those observed in signal/boundary systems: organelle hierarchies in biological cells, ecological niches consisting of generalists and specialists, language communities consisting of speakers sharing a common dialect, political units consisting of allied interest groups, stations in a production line, and so on. In order for the *dgs* to be of interest, the following more specific goals must be achieved:

(i) diverse multi-agent interactions based on tagged signals

(ii) conditional rules that favor "production-line specializations," "recycling" loops, and other signal-boundary interactions

(iii) continuing changes in organization produced by adaptation and evolution.

Grammars offer a step in the right direction insofar as they produce a generated corpus, but the usual grammar is ill-suited for investigating *dynamic* interactions of adaptive agents. However, when tags control signal processing and the passage of signals through boundaries, signal/boundary dynamics can be brought into play in a grammar-like setting. In particular, tagged urns that exchange balls (signals or reactants) in a diffusion-like way can specify the dynamics of general reaction networks constrained by boundaries. (See chapter 4.)

Within a *dgs*, a community of interacting tagged urns can define the bounded reaction network of an agent. Under this provision, coevolution becomes a matter of (i) subjecting the agents to Echo-like requirements, where an agent's replication requires that it collect the "letters" (elementary generators) necessary to copy its structures, while (ii) recombining tags to produce diversity in the exchange of signals and resources. As was emphasized in chapter 6, agents with boundaries and signals that provide enhanced rates of relevant letter collection replicate more rapidly, so their signal/boundary building blocks occur more frequently in succeeding generations. The tags used by the signals and boundaries then become grist for the coevolutionary mill. Recombination comes into play when replicating agents can be "crossed."

It is helpful to re-examine the four interlocking models of signal/boundary interactions discussed in earlier chapters, keeping the points just made in mind. The italicized words in the following four descriptions refer to the five requirements—

building blocks, spatial distribution, programmability, implicit fitness, and internal models—set forth at the end of the preceding chapter.

• **Classifier-system models** (chapter 3) specify the *performance/ credit assignment/ rule discovery* activities of an adaptive agent in a complex adaptive system. Because a classifier system (*cfs*) uses a population of rules (strings), rule discovery can be implemented with a genetic algorithm, using *recombination* to generate new tags. Because the rules can be sequenced, *cfs* models can implement any computer-executable model, supplying *programmability*. In particular, sets of rules can act as *internal models*—subroutines that allow an agent to look ahead and *anticipate* consequences of current actions.

• **Reaction nets** (chapter 4) model the interactions of adaptive agents in collecting and processing resources. The *flows* of resources in a population of agents are handily modeled with reaction nets. When tags are used to define the net, progressive changes in the net can also be generated.

• **Tagged urns** (chapter 7) specify the effect of *boundaries* on the throughput of coupled reactions. Boundaries can be modified by changing the entry and exit tags, thus making it possible to study the coevolution of signals and boundaries. Urn models also lend themselves to mathematical studies using Markov processes. (See chapter 15.)

• **Echo models** (chapter 9) situate agents in a *geometry* and specify an agent's structure in terms of elements that can be obtained by collecting resources available at sites in the geometry. By requiring an agent to collect the resources necessary for replication, an *implicit* fitness is introduced—a fitness that can change according to the context provided by other agents (e.g., resource exchanges). With this provision, the concept of *niche* is explicitly defined in terms of communities of agents.

One reason for employing the *dgs* approach is that all the signal/boundary systems we have examined are generated from simple "alphabets":

s/b system	signals	alphabet
genetics	chromosomal transcripts	nucleotides
molecular biology	proteins	amino acids
ecological niches	resources	C, N, O (and some "trace elements")
Language communities	utterances	phonemes
music	phrases, melodies	notes
political units	memoranda	written alphabet
production lines	raw materials	atoms

In each of the four models just discussed, strings over an alphabet can be used to stand for the basic components: signals, reactants, or billiard balls. The same alphabet can be used to define catalysts, which can be represented as "rule-marked balls" in urn models. (See chapter 8.) By designing the *dgs* with close attention to the universality of the string-processing rules provided, any of the variety of mechanisms and processes associated with signal/boundary systems can be simulated. Then the tags used by tagged urns can define the complex hierarchies of semi-permeable membranes found in signal/boundary systems. Agents then can be defined within the *dgs* as communities of tagged urns.

12.3 An agent-based *dgs*

In a *dgs*, agents are delimited by the outer boundary that contains their signal/boundary hierarchy. A biological cell, for example, can be treated as an agent with a hierarchy of organelles and other structures that define its capabilities. In general, agents must repeatedly execute three basic subroutines: internal signal processing, interaction and migration, and replication. These subroutines must be specified as strings constructed of the generators of the *dgs* framework, then the framework's computation-universality can be exploited to provide progressive variation of the subroutines.

Let us now examine the three basic subroutines in more detail.

Internal signal processing

In the urn-based models we have examined, an agent uses three mechanisms for signal processing: drawing billiard balls at random from the urns, moving drawn balls from one urn to another when their tags satisfy the entry and exit conditions of the urns involved, and pairing drawn balls to represent collisions between balls. In terms of these mechanisms, earlier signal-processing discussions can be summarized as follows. Balls have markings ("colors") that designate specific signals and reaction rules. Because the balls are drawn at random, the effects of the three mechanisms just listed can be described within a particle mechanics, where a random drawing of a pair of balls is the counterpart of a concentration-based random elastic collision. The number of draws per time step corresponds to the collision rate (temperature) in particle mechanics. During a collision-based reaction, the generating elements (letters) are recombined to "recolor" the balls, which then replace the balls that were drawn. (When rule-marked balls are used as catalysts, the overall reaction may involve two pairings.) With these mechanisms, the agent's signal-processing activity at a given time is described by the totality of drawings, movements, and pairings taking place in its defining urn hierarchy at that time.

An agent's signal processing, so defined, corresponds to the signal processing provided by a classifier system, so that the population of signal-marked balls at any time corresponds to the signals on the *cfs* signal list. There is, however, an important difference. In the urn model, several balls may be marked with the same signal. Because of this multiplicity, a message may be processed by the same rule many times, producing many copies of the signal. This multiplicity amounts to a weighting of the signals, the more frequent signals being processed more often. The result is a probabilistic version of a classifier system in which successive *populations* take the place of successive *cfs* signal lists. This probabilistic model retains the computational completeness of a *cfs*, but in a stochastic form that centers on the density of the "answer." (See chapter 15.)

Interaction and migration

For purposes of agent interaction and migration, the whole agent is treated as a single unit, much as if it were a billiard ball. Interactions between agents, then, are initiated by drawing two agents at random from a site. The entry and exit conditions of each agent's outer (exterior) urn determine its possibilities for interaction with the other agent. For example, an exchange of signals is accomplished as follows. Draw a ball from the outer urn of one of the agents. If the ball matches both the exit condition of that agent and the entry condition of the other agent, move the ball to the outer urn of the other agent. The same procedure is used for exchanges within an agent, and is repeated in proportion to the number of balls in the selected inner urn.

Random migration of an agent to a new site is accomplished by drawing an agent at random from a site, then moving it to an adjacent site. That is, the coordinate tag of the agent's outer urn is changed to match the coordinate of the randomly chosen adjacent site.

Replication

Agent replication depends on making a copy of the urn hierarchy that defines the agent. In particular, the agent must provide copies of the entry and exit conditions of the urns in its hierarchy. The agent's fitness then depends on its ability to collect resources that supply the letters used in making these copies. In the *dgs* framework, the incoming letters result from exchanges of strings (resources, signals) between agents, and strings from the site (the non-agent milieu surrounding the agent).

A rate of replication set by resource exchange opens a wide range of coevolutionary possibilities. If some resources are readily available, an agent may retain a simple structure that exploits the elements (letters) supplied by those resources, with a correspondingly high replication rate. At the other extreme, an agent may have a complicated structure that requires many elements, each urn in that structure acting as a specialist in an efficient "production line" dedicated to replication. A diverse array of agents may also "shuffle" and transform resources, forming a niche that offers sustainable replication rates for each of its members. When each agent has a probability of being deleted (e.g., a "half-life"), there is strong selection pressure against agents that replicate slowly or not at all. To enforce conservation of elements, the deleted agent's structural elements can be released to its site, thereby making the elements available to other agents at the site.

To keep replication a simple process, the *dgs* can have an operator at the meta-level that disassembles accumulated strings and then reassembles them into entry and exit conditions for copies of the agent's urns. In the simplest models, a description of the urn hierarchy, compiled at the meta-level, is used to direct the replication. Because the description is at the meta level, replication of the description doesn't require any of the accumulated letters. A more complex model imposes a selective pressure for shorter descriptions by requiring that the description, too, be assembled from accumulated elements. In a still more complicated model, the *dgs* meta-operator is replaced by a set of urn-based coupled reactions. Note that the description of the urn hierarchy, however it is implemented, serves as a "chromosome" for the agent.

12.4 Conglomerate agents

Now consider agents that are *conglomerates*—that is, collections of agents, as described in chapter 5. Conglomerates become especially important when we consider the progressive complication in signals and boundaries (*ontogeny*) that occurs when a multi-celled animal (a *eukaryote*) develops from a single fertilized egg cell. Because all cells in a given animal have the same chromosomes, this process requires signal proteins that turn parts of the chromosome on and off. A similar requirement holds for *dgs* description of a conglomerate. There must be a description that encompasses the diverse possibilities for component agents.

To make this provision, the description must have sections corresponding to different kinds of components. Each section (let's call it an *operon*, appropriating a related biological term) has two parts: (1) a part that gives one or more conditions for signals that turn the operon on/off (called *activation/repression* in molecular genetics) and (2) a description of one of the agents that is a component of the conglomerate. The description of the conglomerate then consists of a set of operons. With this provision, the on/off status of a given operon depends on the signals present in a parent agent at the time of replication. The operons that are on determine which of the conglomerate's agents are present in the offspring. Because the offspring of successive replications "stick together," the result is a progressively more complex conglomerate made up of many different agents, mimicking the ontogeny of a eukaryote.

In tagged-urn terms, the balls present in the parent urns determine which operon-defined urns are present in the offspring. Structural differences between parents and offspring (if there are any differences) depend on differences in their arrays of activated operons; the parents and the offspring have the same chromosome-like description. (This process will be described in more detail in section 13.4.)

If agents' external urns are supplied with a *mating tag* and a *mating condition* (again, in the fashion of the Echo models), agents can be mated and cross-bred within the *dgs* framework. That is, when two agents come into contact, they can mate if the mating tag of each satisfies the mating condition of the other. Typically, the *dgs* mating operator restricts mating to agents that have accumulated enough elements (letters) to replicate. When the agents mate, their descriptions are crossed. The descriptions so modified (in the manner of a genetic algorithm) become the descriptions of the offspring. Note that a mating condition with many #'s allows many different kinds of mates, with the possibility that many offspring will be of low fitness; on the other hand, a condition with few #'s is very selective, which makes finding a mate more difficult but increases the likelihood that the offspring will survive.

12.5 Onward

This chapter has outlined the form of a *dgs* suitable for a unified study of signal/boundary coevolution. Chapter 13 presents a working model based on this outline. Chapter 14 presents a completely defined, overarching *dgs* of this kind.

13 A Dynamic Generated System Model of Ontogeny

13.1 A simple model of ontogeny

What follows is an idealized model, presented within the *dgs* framework, of biological development (ontogeny—see Forgacs and Newman 2005) in an organism with a spore stage. The model doesn't match any particular developmental process, though it could easily be made more realistic. Rather, the model is designed to show the kinds of developmental activities that can be achieved by the interaction of five well-known biological mechanisms: gene (de-)repression, semi-permeable membranes, cell adhesion, catalyzed reactions, and replication through reassembly of collected resources. In addition, it illustrates the ways in which semi-permeable boundaries can generate signaling networks by directing and re-directing diffusion processes (modeled by a billiard-ball mechanics).

The model starts the developmental process with a single inactivated "spore" agent. The "spore" agent is activated when a critical resource (say, "water") is present, at which point the agent begins to collect and process resources. After collecting sufficient resources, the agent replicates. The structure of the offspring is determined by a "chromosome" consisting of two "operons." The whole chromosome is passed on from parent to offspring, but either or both of the operons may be repressed by signals originating from within the replicating parent. Only structures specified by the active (de-repressed) operons in the parent are present in the offspring; this opens up the possibility of structural variation in successive replicants.

In more detail, the "spore" agent consists of a single bounded compartment with a structure determined by operon 1, the first of the two operons. When the critical resource (again, think of "water") appears, the "spore" begins to process resources using the reactions specified by operon 1. The signals resulting from this processing cause operon 1 to be repressed ("turned off"), at the same time that operon 2 is de-repressed ("turned on"). As a result, the offspring that is produced when sufficient resources have been processed has a structure determined by operon 2. As will be explained, once the offspring is produced, it is immediately "engulfed" so that it becomes a compartment inside the parent. Then, as long as the critical resource is present, this newly compartmented agent replicates as a whole, both operon 1 and operon 2 being de-repressed. The replicants stick to each other to form a colony. When the critical resource disappears, the agents in the colony produce free, un-compartmented offspring specified by operon 1. These offspring are in "resting" mode because of the absence of the critical resource (that is, the colony produces "spores"). The whole spore-to-spore cycle is started again when the critical resource reappears.

To keep the model simple, replication is treated at the meta-level as discussed in section 12.3. When an agent processes enough resources to make a copy of the structures specified by its active operons, the offspring is automatically produced. In a more sophisticated model, the replication process would be "spelled out" at the reaction level. Particular resources would be directed to a "reservoir" (another compartment) with the help of tags, then the reservoir would use reactions (specified by the chromosome) to assemble the offspring agent. Note that replication in this more sophisticated model would require additional elements from the reservoir to make copies of the "replication reactions," a requirement that adds an evolutionary "pressure" for simple replication procedures. In an even more sophisticated model, the agent would also have to collect additional resources in order to reassemble them into a copy of the "chromosome." However, as we will see, many ontogenic effects produced by the five basic mechanisms can already be examined in the simpler model.

13.2 Five mechanisms

The five mechanisms underpinning the model—gene (de-) repression, semi-permeable membranes, cell adhesion, catalyzed reactions, and replication by reassembly—can be simplified and formalized as follows.

Activation

Each operon has one or more conditions that, when satisfied by ambient signals within the agent, cause the operon to be activated (de-repressed). The structures specified by the activated operon appear in any offspring of the agent.

Sample notation: ***a&-z1** indicates an operon that is activated when **a** is present and **z1** is not present.

Entry and exit

Each agent boundary has a set of entry and exit conditions that determine what signals can enter or exit through that boundary.

Sample notation: >**a** indicates a boundary that passes signal **a** into the agent's interior; <**a** indicates a boundary that passes **a** from the interior to the exterior.

Adhesion

Each agent has a set of conditions that determine whether or not it adheres to a second agent that comes into contact with it; agents that adhere to each other move as a unit under diffusion. One of the adhesion conditions acts as a "receptor" that must be satisfied by some part of the second agent's structure. The other condition requires that the agent have specified signals present in its interior (so that it can control its "stickiness"—see below).

Sample notation: **@z|y1** indicates an agent that, on contact with a second agent, will adhere to the second agent if, at the time of contact, the first agent has signal **y1** in its interior and the second agent presents signal **z**.

Reaction

Each operon in an agent's description has a part that specifies a set of "catalysts." The catalysts determine which reactions are favored when reactants (e.g., proteins) collide. In effect, the catalysts determine the signal processing in the compartment(s) defined by the operon. The part of the operon that is producing catalysts can be "turned off" (repressed) by signals within the agent. (See section 13.5.)

Sample notation: **/-y1/&a,y/y1** is a catalyst that, when **y1** is not present in the agent, combines **a** and **y** to produce **y1**.

Replication

As in the Echo models, each agent collects resources in a "reservoir" through interactions with the environment and other agents. When the agent has enough elements in its reservoir to copy the structures specified by the active operons, the reservoir sends a "ready" signal (**r**). The "ready" signal, sometimes in combination with other signals, initiates production of the offspring agent. (See below.) In this simple model, signals with appropriate tags are automatically directed to the reservoir, where they are automatically reassembled into offspring structures.

Sample notation: **$a&r** indicates that the agent replicates when the signal **r** indicates that the reservoir contains enough elements to copy the agent's structures and the signal **a** is present.

In models based on these five mechanisms, an agent is specified by a "chromosome" consisting of a string of operons. Each operon selects mechanisms from the set of five to specify the capabilities of a *compartment* (a bounded subunit of the agent, like an organelle). By convention, if an operon doesn't specify at least one entry (exit) condition for the corresponding compartment, all strings pass into (out of) the specified compartment. If the operon has no specification for one of the other mechanisms, that mechanism is inoperative. Thus, an operon without an activation condition is never active, an operon that specifies no reactions has no reactions, and so on.

13.3 Specification of the "spore" model

The "spore" model has only two operons: 1 and 2, defined as follows. (By convention, a letter in these definitions, such as **a** or **y2**, stands for a prefix, so that any string with that prefix satisfies a condition using that letter.)

	Operon 1	Operon 2
Activation (*)	*a&-y1 *-a	*a
Interpretation	(1) is activated when the signal **a** is present and the signal **y1** is absent.	(2) is activated when the signal **a** is present.
Entry, exit (>, <)	>**a** >**x** >**y**	>**a** >**x** >**y1** <**y2**
Interpretation	**a**, **x**, and **y** can enter the compartment specified by (1).	**a**, **x**, and **y1** can enter the compartment specified by (2), and **y2** can exit the compartment
Reaction (/)	**/-y1/a&y/y1**	**/-y2/a&y1/y2**
Interpretation	This reaction catalyst is only produced when **y1** is absent. When the reactants **a** and **y** collide with this catalyst, the product **y1**is formed.	This reaction catalyst is only produced when **y2** is absent. When the reactants **a** and **y1** collide with this catalyst, the product **y2** is formed.
Adhesion (@)	**@a&y2**	*no adhesion*
Interpretation	Operon 1 specifies an agent that adheres to any contacting agent containing **y2**.	

	Operon 1	Operon 2
Replication ($)	**$a&r**	**$a&r**
Interpretation	The agent replicates when signals **a** and **r** are present. Signal **r** is only present when the reservoir contains enough elements to replicate the structures of the active operons. [Recall that an agent has a single merged reservoir for all compartments.]	

Signals with tags **x** and **y2** are automatically directed to the reservoir for re-assembly into agent structures.

With these conventions, the agent's "chromosome" is given by the following string, where ";" separates the two operon descriptions:

***a&-y1*-a>a >x >y <y1/-y1/a&y/y1@a&y2$a&r;*a&-z1> a>x>y1<y2/-y2/a&y1/y2$a&r**.

13.4 Ontogenic sequence

In the ontogenic sequence that follows, [1] designates an agent with the single compartment specified by operon 1, [2] designates an agent with the single compartment specified by operon 2, and [3] designates an agent with outer structure specified by operon 1 and an interior compartment specified by operon 2.

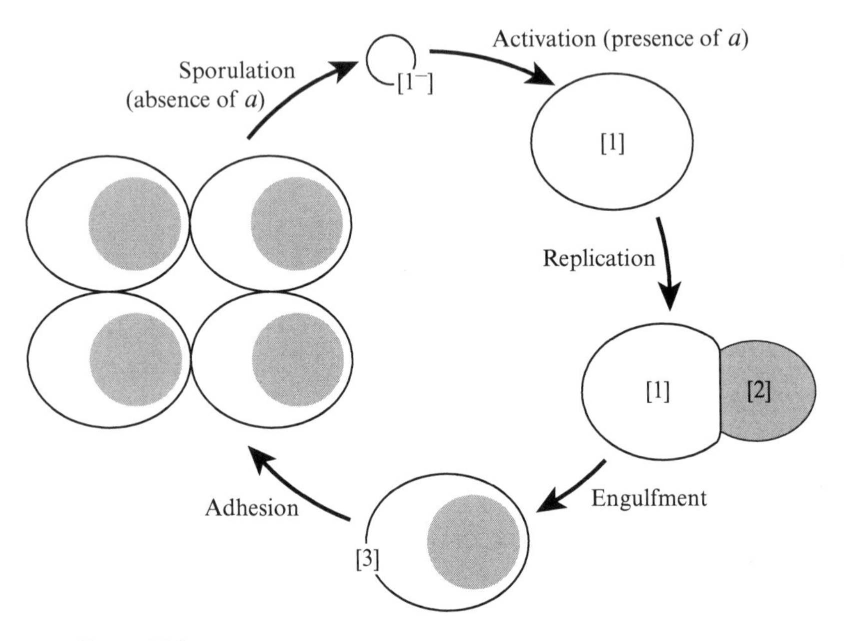

Figure 13.1
A model of ontogeny.

In detail, then, the ontogenic process goes through the following stages:

(1) Inactive agent [1], "spore" mode
In the absence of **a**, the reaction **/a&y/y1** doesn't take place, so [1] is quiescent, though it admits **x** and **y** from the environment as specified by the segment >**x** >**y** in operon 1. Because of the absence of **a**, [1] doesn't adhere to its parent [3] when it is formed, so it is set free of the colony that produced it.

Content (signals/billiard balls)

x, y, a&y/y1

(2) Activated agent [1]

When **a** (re-)appears in the environment, agent [1] starts producing **y1** via the reaction **a&y/y1.** The formation of **y1** deactivates operon 1 via ***a&-y1**, while ***a** activates operon 2. As a result, an offspring of [1] will be of type [2]. (Note again that the chromosomes of [1] and those of [2] are identical; activation only determines the form of the offspring, so that structures already present in [1] (the parent) are not affected by changes in operon activation.)

Content

a, x, y, y1, a&y/y1

(3) Replication of [1]

As specified by **$a&**r, [1] replicates when **a** is present and signal **r** is sent by the reservoir, indicating that [1] has collected enough resources to produce the structures specified by the active operon 2. At that point [2] is produced. Because of [2]'s entry condition <**y1**, [2] contains the reactant **y1** produced by [1]. Agent [2] produces **y2** via the reaction **a&y1/y2**.

Content [1]

a, x, y, y1, a&y/y1

Content [2]

a, y1, y2, a&y1/y2,

(4) Formation of agent type [3] from parent [1] and offspring [2]

The parent agent [1] adheres to its offspring [2] via **@a&y2**, whereas agent [2] has no adhesion condition. This asymmetric adhesion causes [1] to engulf offspring [2]. The result is the formation of an agent of type [3], a compartmented agent with a single reservoir (automatic merger of the reservoirs of [1] and [2]). Operon 1 is reactivated, because **y1** is no longer present, having been transformed to **y2**; operon 2 remains activated.

Content (outer compartment)

a, x, y, y1, a&y/y1

Content (inner compartment)

a, y1, y2, a&y1/y2

(5) Offspring of [3] with resource a present

Because both operons are activated, the offspring of [3] are of type [3]. The adhesion property of [3] is determined by the outer compartment, specified by operon 1. Therefore, because of the presence of **a** and **y2**, **@a&y2** causes the offspring to adhere to the parent, and vice versa, resulting in a colony.

Content (outer compartment)

a, x, y, y1, a&y/y1

Content (inner compartment)

a, y1, y2, a&y1/y2

(6) Offspring of [3] with resource a absent (spore production)
When **a** is no longer available in the environment, the offspring of [3] is an agent of type [1] with a single compartment. Because **a** is absent, offspring [1] doesn't adhere to [3] or vice versa; thus the offspring diffuses independently of the colony. Because the replication requirement **$a&r** isn't satisfied, [1] doesn't reproduce, so it acts as a "spore" until **a** reappears.

Content

x, y, a&y/y1

13.5 Generating the model

The objective of a *dgs* is to spell out all the mechanisms as strings over a single small alphabet. The mechanism identifiers{*, **&**, -, /, >, <, @, **$**, ;}, in combination with the {1,0,#} alphabet, suffice to define all the operon structures employed by the model, as well as all the tags, signals, and reactants. In exploring the effects of recombination, it is sometimes helpful to reduce this alphabet still further to the four-letter alphabet {1,0,#,g}, using g as the initial and the final letter of the encoding of the mechanism identifiers. For example, one set of encodings would be

* is encoded by gg, **&** by g101g, - by g000g, / by g110g, > by g001g, < by g100g, @ by g1111g, **$** by g0011g, and ; by g1g.

Mechanism conditions

Each mechanism involves a condition or multiple conditions that must be satisfied in order for the mechanism to be activated. Any finite string c over the alphabet {1,0,#} can be interpreted as a condition as in a classifier system. If c is terminated by #, then c accepts any signal (string) with a prefix that matches c; otherwise c accepts any signal that matches the non-# letters in c position for position. Thus, as with classifier systems, the condition 111# accepts any signal (string) with the prefix 111, the condition 1#10 accepts only the signals {1110, 1010}, and so on. If c is a condition, then –c is a condition that is satisfied only if there is no signal matching c. If c is a condition and d is a condition, then c&d is a condition that is satisfied only if there is a signal x matching c and there is a signal y (not necessarily distinct from x) matching d.

Conditions, so specified, can also be used as conditions for reaction catalysts. The same alphabet can specify the *actions* (products) of the reaction catalyst. A finite string b over {1,0,#} specifies an *action* when it is the string following the last / in the reaction specification. If b contains no #, then b is added to the set of signals in the urn containing the catalyst. If b contains one or more #'s, a pass-through (see section 2.3) is indicated. The letter at the corresponding position in the signal satisfying the mechanism's (first) condition is inserted in the outgoing signal specified by b. If p is an action and q is an action, then p&q is an action, interpreted as two outgoing signals.

Operon activity

In this model, as in most signal/boundary models, it is critical to set the rates at which operon-directed activities take place. For example, there must be a parameter determining the number of copies produced per time step of, say, an operon-described catalyst x. Here we can use the earlier discussion of reaction networks, in which rates of collision between reactants (balls) determine the rates of reactions. In the present context, we can treat each condition for catalyst production as a target for collisions, just as if the condition were a ball within the urn. That is, when a ball from within the urn strikes the catalyst condition and satisfies it, an additional copy of the catalyst x is placed within the urn.

A self-regulating approach mimicking gene repression is suggestive. Set one of the conditions for production of the catalyst so that the condition is *not* satisfied whenever a catalyst ball x already present in the urn collides with the condition. Then, as the number of balls of type x increases within the urn, more and more collisions occur with this condition, and the operon-directed production of x slows. This kind of self-regulation, as we will see in the next chapter, plays a major role in signal/boundary coevolution.

Operon replication

For purposes of replication, the letters of the alphabet {1,0,#,g} are treated as "elements" that can be reassembled to designate new mechanisms, reactions, or reactants. For example, the reactant designated by 11011, when disassembled, supplies the agent with four copies of element 1, one copy of 0, and no copies of the other elements. An agent obviously benefits by collecting reactants that supply a high proportion of the elements it needs for making copies of its structures. This collection process depends on what resources can pass through the agent's boundaries. In the simple spore model presented above, strings to be used for replication are redirected to the agent's reservoir, wherein the elements are recombined to form the structures specified by the active operons.

13.6 Signal routing in the model

Now we can look at the way boundaries affect the routing of signals in the ontogenic process. The use of urns to represent the bounded compartments provides the starting point.

Each operon in the description specifies one urn, and each urn has a unique id (identity number). The id uses decimal subsection notation (discussed in section 8.1) to indicate the position of the corresponding compartment in the agent's boundary hierarchy. Thus, if agent [3] has id = 244 and [2] is an interior compartment of [3], the urn corresponding to [2] would have id = 244.1. Note that this notation is part of the description of the urn model; it isn't used in the string of operons that define the agent.

The basic execution step of the urn model consists of N random draws from each of the urns belonging to an agent's boundary hierarchy, where a larger N sets a larger collision rate. The resulting collisions determine the activities (movement of signals through the boundaries, signal processing, and the like) within the agent compartment corresponding to the urn. For example, each ball drawn for diffusion between urns is randomly assigned to test either the containing urn's exit condition or else it tests one of the entry conditions of the urns within the containing urn. If the label of the ball drawn satisfies the condition, the ball is moved into (placed in) the corresponding urn; if the condition isn't satisfied, the ball stays in the urn of origin.

One procedure would use two sets of drawings. After the diffusions are determined, a new drawing of N' *pairs* of balls takes place to test for reactive (signal-processing) collisions. If the tags on a drawn pair satisfy the conditions for a reaction, the balls are transformed into balls with labels corresponding to the products of the reaction. The products replace the pair that entered the reaction; if there is no reaction, the pair drawn is placed back in the urn without change. Catalyst production and self-regulating catalyst production (discussed in the preceding section) then also depends on *N*′. When the conditions for activation of catalyst construction are satisfied (by successive collisions) a new catalyst ball added to the urn. In a model in which the number of elements (letters) is conserved, the reservoir must supply the elements required to make a copy of the catalyst.

Thus, in a complete time step, every urn (compartment) undergoes a diffusion step (*N* drawings) and then a reaction step (*N*′ drawings of pairs). The drawings are treated as if they were executed simultaneously, as would be the case in a real multi-agent system. Once the diffusion and the reactions have been carried out, each agent (hierarchy of urns) is tested for readiness to replicate. When an agent replicates, the offspring agent, with properties specified by the parent's active operons, is placed next to the parent. Then the offspring is tested for adhesion to the parent, as determined by the adhesion conditions specified by the operons of the parent and the offspring. Finally, urns that mutually adhere (colonies), as well as isolated urns, are selected at random from each site for migration to adjacent sites. The coordinates of the urns that move are changed to specify the site they now occupy.

13.7 Summary

This model of ontogeny uses the *dgs* format to reformulate five biological mechanisms discussed in earlier chapters: gene (de-) repression, semi-permeable membranes, cell adhesion, catalyzed reactions, and replication through reassembly. With the spore model as a guide, the next chapter presents an overarching *dgs* system for the study of *arbitrary* signal/boundary interactions.

14 A Complete Dynamic Generated System for Signal/Boundary Studies

14.1 Guidelines

A fully defined *dgs* for studying signal/boundary systems requires the following:

a set of *generators* G (an "alphabet")

a *corpus* produced by combining the generators to form strings that are interpreted as programs determining the dynamics of urn-based agents

a set of *meta-operators* for executing mechanisms specific to signal/boundary systems.

The *dgs* is initialized with a population of agents X(0), a specification equivalent to specifying the axioms of a formal deductive system such as Euclidean geometry. Once X(0) has been specified, the *dgs* is autonomous, and the agent-defining programs act on the initial set X(0) of strings to produce new generations X(1), X(2), X(3), ... Note that X(t) must keep track of both the changing urn hierarchies that define the agents and the contents of the urns in the hierarchy.

The steps producing the successive generations X(t) correspond to time steps in the ontogeny model. All urns (agent compartments) undergo a diffusion step (N_1 drawings) and then a collision step (N_2 drawings of pairs). The collision step determines what changes will be produced by reactions; it also determines which meta-operators will act. (See below.) The drawings are treated as simultaneous, much as a classifier system processes many signals in each time step. Choosing a larger N_1 or N_2 increases the intensity of the corresponding activities. The operations of agent replication, adherence between agents, and agent diffusion to new sites are executed at the end of the time step. In the simpler models these latter activities are carried out by meta-operators.

The designs of the agents in the initial population X(0) determine what kind of signal/boundary system is being investigated—ecological niches, cellular semi-permeable membranes, language communities, and so on. X(0) specifies the balls (reactants and reaction rules) contained in the urns defining the agents, so X(0) determines the emergent dynamics resulting from the actions and interactions of the agents. All *dgs* examined here, whatever the signal/boundary system being modeled, have the same fixed set of generators, G, much as a general-purpose computer has a fixed instruction set. The allowed ways of combining the generators (to define "programs") are also the same for all signal/boundary systems. This common defining structure allows rigorous comparisons of different kinds of signal/boundary systems under different initial conditions.

14.2 Overview of *dgs* generators and operators

As in the ontogeny model of chapter 13, the classifier system's alphabet {1,0,#, -, &, /} provides a starting point for the *dgs* generators, sufficing for the strings that specify the tags, signals, and reaction rules of the urn-based agents. Repeating the interpretation used in the ontogeny model, we have the following:

Strings of ones and zeroes specify signals.

is a 'don't care' used in defining conditions for rules.

- is interpreted as "not," so that -10# specifies a condition satisfied only if no signal with prefix 10 is present (in the urn).

& is interpreted as "and," so that the combined condition 10000&01110 requires that both signal 10000 and signal 01110 be present simultaneously (in the urn).

/ is interpreted as THEN, separating the condition part of the rule from the action (output) part.

Thus, the string 01011 & -10# / 00000 designates the reaction rule

IF [signal 01011 is present *and* no signal with prefix 10 is present]

THEN [send signal 00000].

When a rule produces more than one output signal simultaneously, the symbol & is used to separate the specifications of the outgoing signals. For example, the string 10011/100&11 designates the rule

IF [signal 10011 is present]

THEN [send signals 100 and 11].

If a # appears in the output side of a rule, then the corresponding part of the incoming signal is passed through to the output signal. For example, if the rule 100#/0#0 is satisfied by the incoming signal 100**11**, then the rule produces the outgoing signal 0**11**0. With these provisions, the same alphabet can be used to describe reaction networks and entry and exit conditions for urns.

The heart of the *dgs* is the set of generators used to specify urn processing activities—the urn entry and exit conditions and the balls (reaction rules and signals) contained in the urns. To specify urn entry and exit conditions, as in the ontogeny model, the additional generators > and < are required; the set of generators then is

{1, 0, #, -, &, /, >, <}.

As an example, consider an urn with the following specifications.

Entry conditions

>100

>0#

Exit conditions

<111

<10000#

Signal-processing rules

100&00#/10000# (two copies)

1110#/111&0# (one copy)

Signals

100 (three copies)

00000 (one copy)

The operon specifying the structure of this urn is given by the string

>100;>0#;<111;<10000#;100&00#/10000#;1110#/111&0#;100; 00000,

where semicolons separate the components of the specification. The number of copies of the reaction rule in the urn is determined, time step by time step, by the number of "collisions" between signal balls (in the urn) and the corresponding reaction rule part of the operon, as described in the next section.

To replicate this urn, the agent containing the urn must collect resources containing the requisite elements: 18 ones, 32 zeroes, eight #'s, three /'s, two >'s, and two <'s. For example, if the agent's exterior urn admits the signal 100100, that signal can be disassembled to yield two ones and four zeroes. An agent then benefits from collecting resources that supply a high proportion of the elements it needs for making copies of its component urns, a process that depends on which signals (resources) can pass through the urns' boundaries. Thus, in a *dgs* model, the boundaries determine what signals serve as grist for the replication mill.

The *dgs* (but not the operons) must also specify the content of each urn in an urn hierarchy, a requirement similar to using a signal list to keep track of signals processed by the rules in a classifier system. Thus, in the *dgs*, we can think of X(t) containing both a list of urn hierarchies and a coordinated list of urn contents. During urn replication, some of the urn contents of the parent may be passed on to the offspring, the content list being revised accordingly.

In simpler *dgs* models, the details of urn replication and the other activities involving whole urns—urn migration, urn adhesion, and the interaction of urn hierarchies—are controlled by meta-operators. Even when meta-operators are used, the *dgs* must have a description of the specific version of the operators employed by each agent; only then can those particular processes be passed on to offspring. However, where meta-operators are used, resources need not be collected to replicate the descriptions used by the meta-operators.

Three additional generators are used in describing the activities of agents and urns: {* (a prefix for activation conditions), @ (a prefix for an adhesion conditions), and $ (a prefix for replication conditions}. Thus, the complete generator set is

{1, 0, -, &, /, >, <, *, @, $, ;}.

The number of generators for the *dgs* can be reduced, as in the ontogeny model, by following the example of molecular biology, in which chromosomal activities such as operon repression or de-repression are defined using the same four nucleotides that define the rest of the chromosome. As before, we can use the four-letter alphabet {1,0,#,g} to designate the full set of operators {-, &, /, >, <, *, @, $, ;}:

Operator	Representation
-	g000g
&	g101g
/	g110g
>	g001g
<	g100g
*	gg
@	g1111g
$	g0011g
;	g1g

Using these representations, the description

a&-y1-a>a . . . ,

with **a** = **1010101** and **y1** = **1111000**, is encoded as

gg**1010101**g101gg000g**1111000**ggg000g0001g**1010101**. . . .

14.3 Meta-operators and agents

It is possible to design a *dgs* model in which all manipulations, including urn replication, urn adhesion, operon activation, and agent migration, are executed using coupled reactions. But such a model would be even more complicated than, say, a fully defined metabolic network for a biological cell. As a result, large-scale patterns, and possibilities for comparing different signal/boundary systems, would be largely obscured in detail. Fortunately, by using meta-operators, many aspects of signal/boundary coevolution can be examined without going into these complications. In particular, meta-operators can be used to bypass many details associated with replication, adhesion, and migration.

Relevant meta-operators require the following information about each urn h belonging to the urn hierarchy:

the urn containing h

the urns contained in h

the balls contained in urn h.

This information can be organized into four databases:

a *containment list* in which entry h gives the address (position on the list) of the unique urn that contains urn h

a *next-level stack* for each urn on the urn list, in which stack h gives the addresses in the urn list of the urns contained in urn h a *ball stack* for each urn, in which stack h consists of strings describing the balls contained in urn h

an *empty register stack* that records addresses on the urn list that are currently unused (because the corresponding urns have been deleted at some point).

The following are typical (meta-) operations on this database:

Diffusion of a ball (signal)
A ball is chosen at random from the ball stack of urn h and it is tested against either (i) a randomly chosen exit condition of h or (ii) a randomly chosen entry condition of one of the contained urns in the next-level stack of h. If the exit condition chosen in (i) is satisfied by the chosen ball, it is removed from the ball stack of h and is placed in the ball stack of the urn containing h (determined from position h in the containment list). If the entry condition chosen in (ii) is satisfied by the chosen ball, then it is removed from the ball stack of h and is placed in the ball stack of the urn chosen from the next-level stack. The entry and exit conditions thus control the routing of balls between urns.

Creation of a new reaction (signal-processing) rule
Here it is important to distinguish between changes caused by a reaction rule and the creation of a new reaction rule. The creation of a new (copy of a) reaction rule is controlled by the urn's operon. To execute this operation, (i) a reaction-rule segment of one of the operons of urn h is chosen at random and (ii) a ball is chosen at random from the ball stack of h. If the ball satisfies the segment's condition, then a new reaction rule (ball) is added to ball stack h, as in the formation of reac-

tion rules in the ontogeny model. In a more complex model, the meta-operator must obtain, from the agent's reservoir, the elements used to specify the reaction rule represented by the new ball.

Formation of a new urn

The address at the top of the empty register stack, call it j, is assigned to the new urn (and that address is removed from the empty register stack). (If no previously occupied registers are empty, the top of the empty register stack contains, as a new address, the next address after the final address of the current containment list). Then (i) the address of the urn that contains the new urn is stored at address j in the containment list, (ii) the addresses of any urns contained in the new urn are assigned to the jth next-level stack, and (iii) the strings describing the balls contained in the new urn are stored in ball stack j.

Deletion of urn h

Entry h on the urn list is replaced by an 'empty' symbol, and address h is added to the empty register stack.

The complexities avoided by using meta-operators in a *dgs* are clearly illustrated by agent replication. Replication by means of a network of reaction rules would first use catalyst rules to attach tags to selected signals, directing them to the agent's "reservoir" urn. Then further reactions would reassemble the reservoir content into a copy of the agent being replicated. In contrast, a replication meta-operator "sits above" the *dgs*. The complex set of reaction rules is replaced by a meta-operator subroutine that counts the elements available in signals and resources contained within the agent (no reservoir necessary), then reassembles the elements into a replica of the agent's urn hierarchy when enough elements are available. In short, the meta-operator manipulates strings, just as would be done by the agent's reactions, but it produces the end result directly. The meta-operator can easily be extended to encompass other complex aspects of replication, such as recombination of agent descriptions, or passing saved resources to an agent's offspring by manipulation of the ball stacks.

For example, the subroutine executed by the agent replication meta-operator has the following form:

(1) From the description of the active operons of agent x, determine n(x), the numbers of copies of the generators required to produce an offspring of x:

$$n(x) = <n(x,1),n(x,0),n(x,\#),n(x,g)>.$$

(2) Use the next-level stack for x to retrieve the addresses of the urns at the next level of the hierarchy; then use the containment stacks corresponding to those addresses to obtain the next layer of the hierarchy. Repeat this process, forming a list L(x) of all the urns in the hierarchy defining agent x.

(3) Use the addresses in L(x) to access the corresponding ball stacks. Form a list R(x) of all strings with tag **r** in the ball stacks. (Balls with tag **r** are directed to the reservoir of agent x.)

(4) Disassemble the strings in R(x) into generators:

$$r'(t) = <r'(1,t), r'(0,t), r'(\#,t), r'(g,t)>.$$

Record the number of each type of generator.

(5) Determine r(t), the numbers of the generators in the reservoir at t, by adding r'(t) to the reservoir content r(t – 1) from the previous time step:

$r(t) = \langle r(1, t-1), r(0, t-1), r(\#, t-1), r(g, t-1)\rangle + \langle r'(1,t), r'(0,t), r'(\#,t), r'(g,t)\rangle$.

(6) If each of the components of $r(t)$ exceeds each of the components of $n(x)$, then produce an offspring of x, decreasing the numbers in $r(t)$ accordingly.

The movement of an agent from one site to another is similarly simplified by a meta-operator. When an agent moves, the coordinate tags of its component urns are changed to indicate the new location. For example, consider agent 6 at site 8. If it has a two-urn hierarchy, the component urns will be tagged 8.6.1 and 8.6.1.1. A move of agent 6 to site 9 will then change the tags to 9.6.1 and 9.6.1.1. The simplest kind of movement is a move to a randomly selected adjacent site, in effect setting an unconditional migration rate. Even for this simple procedure, manipulating the coordinate tags at the reaction level requires multiple move-relevant reactions. And when the move is conditional on signals it receives from an adjacent target site (a version of chemotaxis), the reaction network becomes still more complicated. However, a meta-operator subroutine can test the signals directly, producing the conditional move almost as easily as the random move.

The other two major activities of urn hierarchies—urn adhesion and operon activation—can be handled similarly.

14.4 The *dgs* corpus

The *dgs* corpus records the progression, over time, of all the details of the agent-based signal/boundary interactions. Generating the corpus is much like deriving theorems from axioms. The system starts from an initial configuration X(0) of urns and their contents. Then the generation of successive parts of corpus is completely determined by the *dgs* operators and meta-operators, which serve the role of rules of deduction in an axiomatic system. As with the corpus of theorems in an axiomatic system, the corpus is produced in a step-by-step fashion from X(0), resulting in X(1), X(2), . . . , X(t), ... In this developing corpus, the urn entry and exit conditions and reactions specified in X(t) roughly correspond to the rules used by a classifier system at a particular time t (remembering that a genetic algorithm makes adaptive changes in *cfs* rules as time elapses). The reactions within the agents, along with changes produced by the meta-operators, modify X(t) to produce X(t + 1).

As in the spore example of the preceding chapter, an agent's structure is described by a string of operons:

Agent description = (operon)(operon) . . . (operon),

Each operon is a string that describes a single urn in the agent's defining urn hierarchy, with the following format:

operon = (coordinate)(operon activation conditions)(urn entry conditions)(urn exit conditions)(adhesion conditions)(reaction catalysts)(replication conditions).

The first substring in the operon specifies the site containing the agent and the position of urn in the agent's hierarchy, as described in the last paragraph of the preceding section. The next substring specifies the conditions under which the structures described by the operon are passed on to the agent's offspring. The remaining substrings describe the other urn mechanisms.

In the *dgs*, then, the agent is presented as a string of strings over the generators {1,0,#,g}. The contents of the urns are also given by strings over the generators. Thus, at any step t, each agent is completely specified as a set of *dgs* strings. Of course, the corpus of strings must be interpreted for the particular signal/boundary system being studied, just as one must supply the meanings for grammatical sentences in a language.

A typical *dgs* interpretation would concern the tropical rainforest discussed in chapter 1. In a *dgs* model of a rainforest, signal strings are interpreted variously as different resources (typically carbon-based compounds) and as reaction-defining strings (e.g., catalysts). Distinct operon-defined agents are interpreted as distinct species. Changes in the operon descriptions (produced by recombination or mutation) yield new kinds of agents (new species). In particular, changes in entry and exit conditions and in signal tags modify the agent's locus in the network of resource flows. In the *dgs*, modification of strings through signal-processing rules (acting as catalysts) then allows combinations of letters (acting as resources) to be transformed and passed from species to species, much as cash is passed from person to person in an economy. In the rainforest, this acquisition of resources is achieved variously by predation, parasitism, symbiosis, exploitation of waste products, and other interactions. Cascades and loops of resource transformation form, helping to slow the exit of resources from the rainforest. Indeed, loops that persist under selection effectively increase the concentrations of useful resources, opening further opportunities for diversification. Even "waste" at the end of these lines offers new opportunities (just as an industrial economy can exploit ecosystem waste products—e.g., guano).

A newly generated agent type can persist only if it taps into the flow of resources in a way that gives it a sustainable replication rate. New agent types that persist then offer further opportunities for interaction, resource transformation, and exchange. When new interactions occur in this way, the complexity of the ecosystem typically increases. The corpus produced by the *dgs*, because it encompasses temporal changes, offers a precise way to explore the effects of different mechanisms on the evolution of an ecosystem.

Language provides a quite different interpretation within the *dgs* framework, using the same basic formalism. In this interpretation, the agents are language users. The entry conditions of an agent's exterior urn become the detectors that react to situational signals (external objects and actions by other agents). The interior reaction rules provide grammar-like processing that organizes strings that will exit the agent's bounding urn. Some of the exiting strings are interpreted, then, as language-like utterances. The agent's operons provide the equivalent of "wired-in" capabilities (such as imitation and shared attention) that can be inherited by offspring. Because the agents must acquire resources in order persist, their language-like capabilities survive only insofar as they augment the agent's ability to collect resources through its language-like interaction with other agents. The adherence operator, under appropriate meta-operators, gives rise to agent communities with the same dialect, much as cells in a metazoan organ interact with some kinds of cells and reject others. The language capabilities of the agents can then evolve under the recombination and selection provided by a genetic algorithm. Again, note that all the formal elements of the *dgs* remain unchanged; only X(0) and the interpretation are changed.

Dgs interpretations for the other signal/boundary systems that have been discussed are constructed in a similar way.

14.5 The generated dynamic

The following life cycle for agents is typical of the dynamic generated by *dgs* operators and meta-operators:

1. *Reactions* Inside each agent, coupled reactions take place in the interior tagged urns.

2. *Diffusion* In step 1 some reactions tag reactants for export, providing for exchanges with other agents; other reactants are directed to the agent's reservoir, where they are disassembled into individual elements (letters) in preparation for agent replication. Operons with conditions satisfied by reactants are activated.

3. *Replication* After an agent's reservoir accumulates enough elements (letters) to make copies of the urns designated by its activated operons, it is paired up with another agent at the site that is ready for replication (if there is such an agent). The descriptions of the paired agents are recombined, and the contents of their reservoirs are used to produce a pair of offspring. Some of the remaining reservoir content (if there is any) is allocated to the offspring.

4. *Adhesion* The offspring are tested for adhesion to parents and neighbors. Some adhesion relations may cause the offspring to be "engulfed" by the parent.

5. *Migration* Some agents at the site are moved to adjacent sites; adhering "colonies" are moved as a unit

6. The next life cycle is initiated by returning to step 1.

There are many variants of this life cycle that fit within the *dgs* format. For example, sites may be supplied with exogenous sources of elements, as in the Echo models; or agents may be supplied with conditional mobility, allowing them to "approach" or "flee" other agents. The variant used depends critically on the questions being asked. Without careful formulation of the questions of interest, it is usually a poor strategy to include additional features just to match a particular natural system.

For questions concerning the origin of boundary hierarchies, it is interesting to "seed" X(0) with condition strings consisting mostly of #'s. These "seeds" are the counterparts of semi-permeable membranes that pass most signals. When a recombination operator is included in the *dgs* repertoire, successive generations yield more restrictive entry and exit conditions, often resulting in cascades and hierarchies. (See chapter 6.) As the hierarchy develops, there is a race between incorporating urns with simple conditions (which replicate more easily but are less efficient at concentrating resources) and incorporating highly selective urns (which require more resources for replication). A default hierarchy—formally a q-morphism (Holland, Holyoak, Nisbett, and Thagard 1986)—is often an effective compromise, offering high selectivity where it is needed while providing simple but less efficient conditions in less critical situations.

Several other questions can be investigated along similar lines. For example, when does early fixation of a component structure outweigh the increased efficiency that can be obtained through further variation and "fine-tuning"? Early fixation makes it easier to employ the fixed component as a "standard" building block for higher-level structures, at a cost of losing possible increases in component efficiency. The Krebs cycle was fixed so early in evolutionary history that it serves as a standard building block for all aerobic organisms, but it is known that the Krebs cycle could be made more efficient. Nevertheless, there was clearly a selective advantage to early standardization in this case.

In studying the balance between fixation and fine-tuning, we come to a question that is implicit in Adam Smith's example of the transition from generalist craftsmen to lines of specialists. How does an initially homogeneous system discover and implement the boundaries and coordinating signals that provide successive stages in a "production line"? Within the *dgs* framework, this dynamic depends on the modification of the tags used by conditions, signals, and resources. Appropriate tags can implement recycling, as in a reaction network, giving rise to the "multiplier effects" (Samuelson and Nordhaus 2009) that favor departures from homogeneous mixing.

14.6 What now?

This chapter has spelled out the use of the *dgs* framework to develop exploratory models. As was discussed in chapter 2, exploratory models come into their own when it is difficult to formulate a plausible conjecture about the origin of some pervasive observation. When that is the case, it is worthwhile to examine ways in which mechanisms observed in similar contexts combine to yield similar observations. A *dgs* provides ways of defining candidate mechanisms, allowing extensive and rigorous exploration of phenomena generated by the mechanisms. The computer-executable aspect of a *dgs* is particularly important when the mechanisms interact in conditional, non-linear ways, as is typical of signal/boundary systems. If the phenomena of interest emerge, then we immediately have an existence proof that the mechanisms can produce the phenomena. But even if the phenomena do emerge, this in no way proves that the observed signal/boundary system produces the phenomena in that way. The model simply shows that it *could* happen that way, yielding a plausible hypothesis. Then an examination of the exploratory model's mechanisms will suggest where to look experimentally to test the hypothesis. Knowing where to look can be a critical step for signal/boundary systems, for which the possibilities for observation seem endless.

For some signal/boundary systems, alternative explanations of observations can be generated by different mechanisms within the *dgs* framework. This existence of alternatives is the counterpart of the well-known fact that many different sets of axioms can produce the corpus of theorems known as Euclidean geometry. Such axiom systems are called *formally equivalent*. If the proposed mechanisms of *dgs* are thought of as axioms, then arriving at the same set of observations using different mechanisms is similar to formal equivalence. However, formally equivalent systems are not equivalent in all senses; they may differ considerably in their intuitive efficacy in answering different questions. More specifically, two different formally equivalent axiom systems may offer quite distinct paths for getting to a particular theorem. Gödel showed that there are axiom systems for Euclidean geometry wherein the shortest proof of the Pythagorean Theorem exceeds any prescribed number of steps (Tarski, Mostowski, and Robinson 2010). Because the Pythagorean Theorem is central to our understanding of Euclidean geometry, there would be a very different "feel" to the geometry if that theorem were not easily attained. Formal equivalence, then, doesn't imply explanatory equivalence. This point applies with equal force to a *dgs*, because the choice of mechanisms and initial corpus X(0) amount to selecting axioms for a *dgs*.

With this caveat about formal equivalence in mind, what mathematical tools will help us in exploring signal/boundary questions? The objective is to use the *dgs* framework to extend locally relevant mathematics to broad classes of signal/boundary problems. Both reaction networks and formal grammars fit well within the *dgs* framework, but there are some substantial barriers to using the accompanying bodies of mathematics in the broader *dgs* framework. To date, research on networks has concentrated on simple, unconditional rules for generating networks, but questions about signal/boundary systems center on networks generated by the coevolution of signals and boundaries. As a result, signal/boundary networks are continually changing, posing substantial challenges to the snapshot-like approach of standard network theory. There is a different barrier to applying the mathematics of formal grammars to signal/boundary systems. The usual studies of formal grammars do not provide a way to examine time-dependent changes within the generated corpus. For example, formal grammars for the study of language rarely examine the effect of time on interactions of language-using agents (though there are exceptions; see, e.g., Steels and Kaplan 2002). Studying the evolution of a grammar is like labeling each theorem in an axiom system with the length of its shortest proof. Then we could examine changes in the order of generation ("evolution") under different mechanisms.

At this point, questions about mathematical extensions of the *dgs* arise. In the *dgs* framework, such extensions require that the mathematics encompass the use of tagged urns and tagged signals, with emphasis on conditional rules that favor cascades ("production lines") and loops ("recycling"), on diverse multi-agent interactions using signals based on shared motifs (building blocks), and on adaptive coevolutionary changes in tags and tag-based interactions. As we will see, the tagged-urn approach opens the way to examining the conditional probabilities of urn entry and exit. The entry condition on an urn sets a conditional probability of entry:

$$P(\text{entry}|b) = p,$$

where b is the required tag and p is the probability that a certain ball will contact that entry condition. Using Markov processes, a mathematics based on conditional probabilities, this basic idea can be extended to more complex *dgs* mechanisms and interactions. The next chapter examines these possibilities.

15 Mathematical Models of Generated Structures

15.1 Prologue

Conditional actions have been discussed at several points. Two examples that have been emphasized are the conditional nature of catalytic effects in biological cells and the role of IF/THEN instructions in universal computers. Conditional actions, above all, play a critical role in the signal processing of *cas* agents. But conditionals pose substantial difficulties for a mathematical approach because they generate non-additive interactions. There is, however, one established body of mathematics that deals with conditionals: Markov processes. This chapter introduces Markov processes as a way of studying tagged-urn models of signal/boundary systems. The definition of a Markov process is elementary—the exposition that follows makes no requirements beyond high school mathematics—but the arguments sometimes require close attention.

15.2 A simple urn-based Markov process

As an introduction to the relation between urn models and Markov processes, consider a system that consists of just two urns and two balls. There are only three possible configurations. The two balls can be in one urn, they can be in the other urn, or there can be one ball in each urn. These configurations can be represented visually as follows:

{|oo||, ||oo|, |o||o|}.

Now let the dynamic of this system be generated by the following "diffusion" rule. Pick a ball at random and transfer it to the other urn. It is easy to see that configurations |oo|| and ||oo| can go only to configuration |o|o|, while configuration |o||o| can go to either |oo|| or ||oo|, depending on which ball is selected. In the latter transition, because the ball (and hence the urn containing it) is picked at random, the two options |oo|| and ||oo| are equally likely, each occurring with probability 1/2.

The transitions induced by the random selection of balls can be specified by conditional probabilities. To simplify the notation, index the configurations {|oo||, ||oo|, |o||o|} as {1,2,3} in the order given, so that configuration |oo|| is represented as 1 and so on. Then we can write P(1|3)—read "the probability of 1 given 3"—as the conditional probability of going, in one step, from configuration 3 (|o||o|) to configuration 1 (|oo||). The set of all one-step possibilities for the system of two urns and two balls is given by the following array:

$P(1|1) = 0 \quad P(2|1) = 0 \quad P(3|1) = 1,$

$P(1|2) = 0 \quad P(2|2) = 0 \quad P(3|2) = 1,$

$P(1|3) = 1/2 \quad P(2|3) = 1/2 \quad P(3|3) = 0.$

This array is the *Markov matrix*, M, for the dynamic of the system. The Markov matrix specifies the basic operation of the system (random diffusion in this case) and doesn't change over time.

After the first draw, we can make another random draw from the new configuration, and we can again calculate the probability of each configuration. The process can be repeated, the probability of each configuration changing after each draw. Thus, starting from an initial configuration, we can assign a definite probability of each possible configuration after t draws, specified by a vector X(t) having one entry (the probability) for each configuration. That is, in this case,

$$X(t) = (P(1,t),P(2,t),P(3,t)),$$

where P(j,t) is the probability that the system has configuration j at time t. For example, X(t) = (0,1,0) indicates that the system is certainly (probability 1) in state 2 at time t, whereas X(t) = (1/2, 1/2, 0) indicates that the system is equally likely to be in state 1 or in state 2 at time t. Note that P(j,t), the probability of being in configuration j at time t, is quite different from P(j|i)), the probability of going from configuration i to configuration j in one time step.

The Markov matrix can be used to calculate the changes in X(t) as time elapses. If P(i,t) is the probability of being in configuration i at time t, then P(j|i)P(i,t) is the probability that j will occur on the next if the draw is made from configuration i. Given the probabilities X(t) for each configuration at time t, the overall probability P(j, t + 1) of being in configuration j at time t + 1 is given by the sum of the probabilities of getting to j from each of the configurations at time t,

$$P(j, t + 1) = P(j|1)P(1,t) + P(j|2)P(2,t) + P(j|3)P(3,t).$$

These sums, yielding the next-state probabilities, follow exactly from the rule for multiplying a vector X by a matrix M. To get to P(j, t + 1), multiply each component X(i,t) of X(t) by the conditional probability P(j|i) at position j in column i of the M, and sum the products. That is, multiply the row vector X by the column vector i of M, as in standard matrix multiplication. Using the earlier example, let $X(t) = (1/2, 0, 1/2)$, so the probability that two balls are in the first urn is 1/2 and the probability that there is one ball in the first urn and one ball in the second urn also is 1/2. Then, using M for that case, we have

$$\begin{aligned} P(1, t+1) &= P(1|1)P(1,t) + P(1|2)P(2,t) + P(1|3)P(3,t) \\ &= 0 \times (1/2) + 0 \times (1/2) + 1/2 \times (1/2) \\ &= 1/4, \end{aligned}$$

with similar calculations for $P(2, t+1)$ and $P(3, t+1)$ using columns 2 and 3 of M. Thus, multiplication by M transforms $X(t)$ into $X(t+1)$; in short,

$$X(t+1) = X(t)M.$$

Because $X(t+2) = X(t+1)M$ for any time t, ordinary matrix multiplication gives

$$X(t+2) = X(t+1)M = (X(t)M)M = X(t)M^2,$$

or, more generally,

$$X(t+T) = X(t)M^T.$$

Here we see an early advantage of the Markov formalism. By simply raising M to power T, we can follow the dynamics into the indefinite future.

15.3 More complex urn-based Markov processes

Now let us consider a more extensive system, with k different colors of balls, n urns with entry and exit tags, and reactions between balls that collide. Let *N* be the total number of all balls of all colors.

Unconstrained diffusion

To implement diffusion, select a ball at random from among the N balls, then move it to another urn, as in the two-urn example. Let the total number of balls of color i (in all urns) be $N(\mathrm{i})$, so that

$$N = N(1) + N(2) + \cdots + N(\mathrm{k}).$$

Further, let $N(\mathrm{h,i,t})$ be the number of balls of color i in urn h at step t. When ball i selected from urn h is moved, urn h loses one ball,

$$N(\mathrm{h, i, t} + 1) = N(\mathrm{h,i,t}) - 1,$$

while the destination urn, h*, gains one ball,

$$N(\mathrm{h^*, i, t} + 1) = N(\mathrm{h^*,i,t}) + 1).$$

These equations can be restated in terms of probabilities, obtaining the updated proportions of ball color i in urns h and h* by dividing through by N:

$$\begin{aligned} \mathrm{P(h, i, t + 1)} &= N(\mathrm{h, i, t} + 1)/N \\ &= N(\mathrm{h,i,t})/N - 1/N = \mathrm{P(h,i,t)} - 1/N, \end{aligned}$$

$$\begin{aligned} \mathrm{P(h^*, i, t + 1)} &= N(\mathrm{h^*, i, t} + 1)/N = N(\mathrm{h^*,i,t})/N + 1/N \\ &= \mathrm{P(h^*,i,t)} + 1/N. \end{aligned}$$

Each distinct arrangement of balls in the urns has a well-defined probability X(t) of occurring at time t, as in the example with two urns. A simple derivation (Feller 1968) shows that the number of ways of arranging x *identical* balls in n urns is

$$\mathrm{C[x,n]} = [\mathrm{x + n} - 1]!/[\mathrm{x} - 1]![\mathrm{n}]!.$$

When there are k different colors, and $N(\mathrm{i})$ balls of color i, the total number of arrangements, v, is the product of the number of arrangements for each color:

$$\mathrm{v} = \mathrm{C}[N(1),\mathrm{n}]\mathrm{C}[N(2),\mathrm{n}] \cdots \mathrm{C}[N(\mathrm{k}),\mathrm{n}].$$

Each of these v arrangements has a probability (perhaps 0) of occurring at time t, recorded by the vector X(t) with v components. X is a *probability vector* (or finite probability distribution) because each component has a value between 0 and 1 and the sum of all its components is 1.

Just as in the earlier example, a Markov matrix (call it D) can be used to specify the conditional probabilities of going from one arrangement to another under a random draw. In this case, D is a v-by-v matrix, the entry at location (i, h) giving the conditional probability P(h|i) of going from arrangement i to arrangement h. D thus defines the dynamic of this multi-urn, multi-color diffusion process.

Though v can be quite large, even for modest numbers of balls, colors, and urns, the dynamic specified by D can still be calculated because of D's special properties. On a single diffusion time step, only one ball is moved from any specific arrangement h. Accordingly, only a few arrangements among the large number possible are accessible from h—those that differ from h by one more ball in one urn and one less ball in another. Thus, (h,i) of D has an entry P(i|h) = 0 for each arrangement i that is not accessible from h. Because most arrangements aren't accessible from any given arrangement, each row of the matrix D will consist mostly of 0's. In the usual terminology, the matrix D is *sparse* (Birkoff and MacLane 2008). There are powerful, fast algorithms for multiplying sparse matrices—the number of steps in the calculation depends only on the number of non-zero entries, not on the number of entries v in each row of the matrix. For that reason, a large v isn't as daunting as it might first appear when it comes to calculating properties of the corresponding signal/boundary system.

Constrained diffusion

So far we have dealt only with urns having no entry and exit conditions. How do the entry and exit conditions come into play? First, the random draw must be extended to allow for "collisions with gates," where each gate is an entry or an exit from the urn. With this extension, the draw of a ball as a candidate for diffusion = requires the execution of a subroutine:

(1) Select a ball at random from all the balls at the site, noting the color i of the ball and index h of the urn containing the ball.

(2) Randomly select either one of the exit conditions of urn h or else an entry conditions of one of the urns contained by h at the next level down in the urn hierarchy (using the *contained urn* stack described in the preceding chapter).

(3) If the ball chosen in step 1 matches the condition selected in step 2, move the ball to the urn selected in step 2; otherwise make no changes.

The net effect of this subroutine is to change the conditional probabilities in the matrix D. That is, there is a probability P(i) of picking ball i (from all balls in all urns), which determines the urn index (h) and the probability P(x|i) of picking a particular condition x from those associated with urn h. If i satisfies condition x, then the arrangement will change from the starting arrangement (call it a1) to the arrangement obtained by moving the ball to the destination urn (call it a2). The probability of this change is

$$P(a2|a1) = P(x|i)P(i),$$

where P(a2|a1) = 0 if ball i doesn't satisfy condition x. This approach, thus, uses urn entry and exit conditions to determine, for each a1 and each a2, the entry P(a2|a1) at row a1 and column a2 of the Markov matrix D.

Reactions

Reactions in urn models, determined by random collisions of balls within an urn, can be interpreted as changing the colors of the two balls that react. Assume that a reaction table gives the colored pairs that result from for each of the k^2 possible pairs of the k colored balls. The model then uses the following subroutine to determine the reactions that actually take place:

(1) Select a ball at random from all the balls in the agent-defining urn hierarchy.

(2) Determine the urn h containing that ball and select a second ball from that urn; if there is no other ball in urn h, no reaction takes place.

(3) From the reaction table, determine the reaction and replace the selected pair with the pair determined from the table.

Under this subroutine, most reactions take place in urns that contain a high proportion of the total balls in the agent, as would be expected in a billiard-ball chemistry. (In a more sophisticated model, the reaction table would be generated by the collisions of reactant balls with catalyst balls, and there could be several distinct outcomes for a given collision, each with its own conditional probability.) As with diffusion, the object is to determine what changes in an arrangement can occur when this stochastic subroutine is executed.

A simple example illustrates the form of a Markov matrix R that specifies the rearrangements that can occur under the reaction subroutine. Let there be only two urns and only two colors, black (b) and white (w). Then there are ten distinct arrangements of two balls in two urns (using the notation for occupancy that was used in section 15.2):

{|bb||, |wb||, |ww||, ||bb|, ||wb|, ||ww|, |b||b|, |w||b|, |b||w|, |w||w|}.

Index the ten arrangements with the numbers {1, 2, . . . , 10}. Let the reaction table be given by the following rules:

b + b => w + b,

w + b => w + w,

w + w => b + b.

Because reactions can take place only when there are two balls in the same urn, only the first six arrangements can be changed by a reaction.

For the reaction w + x => y + z, the rate of production of y and z depends on (i) the probability that w will encounter x and (ii) the probability that a reaction actually takes place when x and y collide. (The latter probability corresponds to the reaction rate, which may be zero.) If the portions (concentrations) of w and x in an urn are p_w and p_x, and w is drawn first, the probability that w will encounter x is $p_w p_x$. Probability ii, the *forward reaction rate*, is given by a reaction-specific constant: $r_{wx|yz}$. These two probabilities, when multiplied, give the probability $r_{wx|yz}$ $p_w p_x$ that a collision of w and x will yield the results y and z. To keep the exposition simple, the rest of this example assumes that $r_{wx|yz} = 1$, so that the reaction always takes place when two reactants encounter each other.

With these provisions, the reaction dynamic can be specified by a 10 × 10 Markov matrix R that gives, for each arrangement, the conditional probability of going to one of the other arrangements. For example, let the current arrangement be |bb|| (state index 1), and let one of the two urns be chosen at random. Then the probability that the empty urn will be chosen is 1/2, and the probability that the occupied urn is chosen is also 1/2. If the empty urn is chosen, no reaction is possible and the arrangement is unchanged, so bb|| goes to bb||; that is, entry 1 in row 1 of R is 1/2. If the occupied urn is chosen, the reaction causes the arrangement to change to |wb|| (state index 2), so entry 2 in row 1 is 1/2. No other reactions are possible, so row 1 of matrix R, corresponding to the current state |bb||, is

1/2 1/2 0 0 0 0 0 0 0 0,

where the zeroes indicate that there are no other outcomes possible under the reaction procedure.

Consider next the case in which the system has arrangement |wb|| (state index 2). Then a reaction can take the system from state 2 to state 3 (|ww||,), so that row 2 of the matrix is

0 1/2 1/2 0 0 0 0 0 0 0.

The rest of the matrix R is filled out in a similar fashion.

In the general case of k colors and n urns, the number of distinct arrangements of the balls is the same for both matrix D (diffusion) and the matrix R (reaction). However, the reaction procedure obviously produces entries in the matrix R that are quite different from the corresponding entries in the matrix D. The entry (i,j) in the matrix R gives the probability of going from arrangement i to arrangement j under the possible reactions. The reaction w + x => y + z contributes the probability $r_{wx|yz}\ p_w p_x$, where p_w and p_x are the probabilities of drawing reactants w and x from a randomly selected urn in arrangement i and $r_{wx|yz}$ is the probability that w and x will react to form products y and z. Replacing the balls w and x with balls y and z results in arrangement j. Summing over all possible reactions that can yield arrangement j from arrangement I gives the entry (i, j) of R. With R so defined, the change in the state vector X(t) produced by the possible reactions is obtained by matrix multiplication, as for diffusion, but with R replacing D. That is,

$$X(t + 1) = X(t)R.$$

As with diffusion, the change produced by T successive reactions is obtained by raising the matrix R to the power T:

$$X(t + T) = X(t)R^T.$$

Diffusion/reaction systems

In thinking about the combined effect of diffusion and reaction, note that even if the initial state is one with no reactions possible (e.g., |b||b| in the earlier example) diffusion will eventually place two reactants in a single urn, making subsequent reactions possible. The combined effect of a diffusion and a subsequent reaction is obtained by multiplying the corresponding conditional probabilities. For example, the conditional probability of diffusion from |b||b| (7) to |bb|| (1) followed by the reaction b + b = w + b yielding |wb|| (2) is given by

$$P(2|1)P(1|7).$$

A quick check shows that the matrix of all possible outcomes of a diffusion followed by a reaction is simply the product of the matrices D and R,

$$X(t + 2) = X(t + 1)R = (X(t)D)R = X(t)DR.$$

More generally, for T1 executions of the diffusion procedure followed by T2 executions of the reaction procedure, the state is

$$X(t + T1 + T2) = X(t)D^{T1}R^{T2}.$$

The product matrix $U = D^{T1}R^{T2}$ is itself a Markov matrix describing the corresponding urn-based diffusion/reaction procedure.

At this point a major theorem about Markov processes can be brought to bear. The theorem applies when, eventually, any arrangement of colored balls in the urns can occur through some combination of diffusions and reactions. That is, for a high enough power of the Markov matrix U, any of the arrangements used to label the rows and columns of the matrix can occur with non-zero probability. A Markov matrix that satisfies this condition (and the process it generates) is called *regular.* Most signal/boundary systems satisfy the regularity requirement, unless there are isolated subsystems in which balls cannot diffuse between subsystems. (There are some technical points here; they are covered nicely in Kemeny and Snell 1976.) The ubiquitous diffusion-like processes in signal/boundary systems make isolated s/b systems rare outside the laboratory. When isolated subsystems are encountered, they can be studied individually.

When a regular matrix U describes the diffusion/reaction system, it can be proved that there is a state X^*, called a *fixed point* or equilibrium, such that $X^* = X^*U$ (Kemeny 1976). That is, after repeated applications of the diffusion/reaction procedure, the system reaches a "steady-state" distribution in which each arrangement occurs with a fixed probability (much like the probability of drawing a black ball from an urn with a fixed number of black and white balls). Furthermore, for any initial arrangement X(0), and for T sufficiently large,

$$X(0)U^{T} \sim X^*.$$

By squaring U, then squaring the result, and so on, producing U, U^2, $(U^2)^2$, . . . , a high power of U can be quickly obtained, allowing a quick approximation of X*. Knowing X*, we then know which urns will have high concentrations of selected ball colors when transients in the signal/boundary system "settle out." That is, we can find out where various reactants and signals are likely to be concentrated in the agent's boundary hierarchy. We can also see the long-term effects of different entry and exit conditions and different catalysts by varying the corresponding entries in the matrices D and R. Note, however, each matrix D represents only a particular set of entry and exit conditions, and each matrix R represents only a particular set of catalysts. To examine the effects of adaptive modification of tags and conditions, we would have to look at the *sequence* of matrices generated by the adaptive changes.

15.4 Relation to reaction networks and grammars

The information conveyed by a diffusion/reaction Markov matrix U can be presented in the form of a network. First, assign a network node to each possible arrangement of balls in the urns. Then, for each pair of nodes (h,i) in the network, place a directed connection from h to i just in case $P(i|h) > 0$ in the matrix U. If each such connection is labeled with the corresponding conditional probability, $P(i|h)$, the Markov process can be reconstructed from the network, and vice versa.

It is interesting, now, to look at directed paths through the network so generated. First of all, there can be directed paths leading from node j to node k, without any directed path leading back from node k to j. (Visually, these states belong to tree-like directed paths leading into, but not out of, the connected part of the network.) Thus, there can be initial states that can never be re-achieved—so-called *transient* states. There are also more complicated networks that can be divided into two disjoint subsets of nodes, V1 and V2, where every edge from a node in V1 must go to a node in V2 and vice versa. That is, in *every* path in the network, successive nodes *alternate* between nodes drawn from V1 and V2. For the corresponding Markov process, this property means that nodes in one subset can occur only on even time steps, whereas nodes in the other subset can only occur on odd time steps. The Markov process still has an "equilibrium," but it is "periodic," alternating between the two probability vectors X1* and X2* associated with the two subsets. (For details, see Kemeny and Snell 1976.) Fortunately, the diffusion processes and the recirculation present in most signal/boundary systems "erase" complications such as these "periodic" subsets.

Markov-generated networks correspond directly to the reaction networks discussed earlier. The vector of reactant concentrations at any time t corresponds to one of the arrangements of the Markov process. Then the reaction equation

$$p_c = p_d = r_{ab|cd}\, p_a p_b,$$

using the forward reaction rate, $r_{ab|cd}$, provides the information for filling in the conditional probabilities in the Markov matrix. The effects of modifying boundaries (membranes) are reflected in modifications of the conditional probabilities assigned to entries in the matrix or to edges of the network.

The *regularity condition*, the concept of *network communities* (Newman, Barabasi, and Watts 2006), the concept of *niche*, and the concept of *agent* are all closely related to the recirculation provided by closed loops in a network. That relation, in turn, brings mechanisms that provide recirculation under coevolution front and center, suggesting the organization of relevant Markov matrices. The matrices let us compare snapshots of the coevolutionary process, but a *dgs* framework is required to study coevolutionary changes in the matrices. The next chapter recapitulates the coevolutionary mechanisms, showing their roles in the overall *dgs* framework.

16 A Short Version of the Whole

This book has explored the origin and elaboration of interacting signals and boundaries in various complex adaptive systems, ranging from biological cells (membrane hierarchies) and ecosystems (niches) to markets (equity and derivative exchanges) and governments (departmental hierarchies). The fact that signal/boundary hierarchies are a common feature of all these systems suggests the possibility of a general, overarching explanation of many *cas* properties. The framework proposed here—dynamic (finitely) generated systems—centers on mechanisms that generate signal/boundary hierarchies.

16.1 Agents

All complex adaptive systems exhibit obvious internal boundaries that divide the *cas* into a diverse array of semi-autonomous subsystems called *agents*. Each agent has a "program" that guides its interactions with other agents and other parts of its environment. To provide a common framework for the programs used by different kinds of *cas* agents, the book concentrates on classifier systems for the following reasons:

- Classifier systems use conditional (If/Then) signal-processing rules, thus capturing the conditional interactions characteristic of signal/boundary systems.
- A classifier system executes many signal-processing rules simultaneously without incurring problems of consistency, thus capturing the simultaneous interactions typical of complex adaptive systems.
- Classifier systems are computationally complete, so any computational procedure (e.g., any agent strategy) can be presented by a set of classifier-system signal-processing rules.
- Tags in a classifier system play the role of active sites, motifs, and the like in different complex adaptive systems, allowing direct comparisons of tag-mediated signal/boundary networks in different complex adaptive systems.
- Classifier systems are specifically designed for use with genetic algorithms, so that the coevolution of signals and boundaries can be generated and studied.

Whereas classifier systems handle the signal-processing aspect of signal/boundary systems, boundaries are specified indirectly, via tags. Tagged urns (chapter 7), a generalized version of the urn models used in probability theory, offer a direct approach to boundaries. Different colors of balls in the urns represent different tagged signals (resources), and the balls can be moved randomly from urn to urn in diffusive fashion, allowing signals to be "broadcast." Boundaries with differing permeabilities to balls with different tags can be directly represented by assigning different entry and exit conditions to different urns. These entry and exit conditions are defined using the same notation used to define conditions for classifier rules. Reactions between balls in the same urn are provided using a "billiard-ball chemistry" (chapter 3). In more detail, tagged urns bring to the fore several characteristic features of signal/boundary systems:

- *Compartmented populations* Using probabilities to designate the concentrations of different colors of balls in an urn allows calculations relevant to the multiplicity of signals in a typical *cas* compartment; urns arrayed hierarchically mimic a hierarchical arrangement of *cas* compartments.
- *Diffusion through semi-permeable membranes* Moving balls from urn to urn, subject to entry and exit conditions, models the effect of diffusion through semi-permeable membranes and the counterparts of constrained diffusion in other signal/boundary systems.
- *Conditional action* Using conditional probabilities to represent reaction rules paves the road for using Markov processes to study signal processing in hierarchical arrays of semi-permeable membranes.
- *Simultaneous processing* Repeated random draws of pairs of colored balls model the effects of simultaneous pairwise reactions occurring in production lines, recycling, niches, and the like.
- *Coevolution* Using a genetic algorithm to modify signal tags and entry and exit conditions offers a direct approach to adaptive, coevolutionary changes that modify boundary permeability.

The foregoing makes it clear that conditional rules, reacting to signals on the basis of tags, play critical roles both in classifier systems and in systems of tagged urns. The two systems can be linked by coding signals and reactants as binary strings, so in both cases conditions can be defined using #'s ("don't cares"). Cross-breeding of tags provides grist for the evolutionary mill, with emphasis on combinations of building blocks for tags that have previously been exploited by the signal/boundary system. For example, a single-point crossover between two conditions typically yields one offspring condition that is more general (has more #'s) and another that is less general. Because complex adaptive systems are organized around populations, both offspring options can be explored without abandoning advantages already attained by the parents. As a result, succeeding generations make greater use of tags that offer additional advantages. In other words, progressive cross-breeding explores the levels between "craftsman" generalists and "production-line" specialists in the milieu of signals and resources available to the agents.

16.2 Niches

The concept of niche has been used to discuss quite different signal/boundary systems, including ecological niches, market niches, personal niches, skills niches, and technological niches. When suitably generalized to multi-agent interactions, it offers a way of comparing these different signal/boundary systems.

The mammalian immune system (chapter 9) provides a well-studied example of the role of signals, tags, and boundaries in delimiting a niche concerned with controlling invading pathogens. The invaders are identified by characteristics of their surface molecules (tags) that distinguish them from body cells and normal resident bacteria. (The problem is complicated because a single mammal is in itself a complex ecosystem, with hundreds of species of resident bacteria co-existing with a large array of body cells. Typically the total number of beneficial, or at least harmless, resident bacteria outnumbers the number of body cells by a factor of 10 or more.) Once identified, the invaders are sequestered in special compartments (vacuoles), where they are dismantled by enzymes keyed to their tags. However, that is only one stage of the evolutionary story. Pathogenic bacteria have evolved signals to redirect the immune system's activities, even converting a cell's interior compartments into safe havens for their reproduction. So the "arms race" continues. Body cells that manage to neutralize the invaders actually flourish in the niche provided by the invaders.

A rather different example of niche formation is provided by the development of a eukaryote from a single fertilized egg. Starting from the fertilized egg, cell division yields an ever larger colony of cells, the cells in the interior experiencing higher concentrations of some proteins than those on the exterior. Differential concentrations, along with induced signals, cause different sets of genes to be repressed or de-repressed in cells at different locations in the growing colony. As a result of this differentiation, cells at different locations in the colony construct different adhesion molecules for their bounding membranes; they also produce new signaling molecules, causing still further differentiation. (See chapter 13.) The development of the mammalian eye is a case in point. Cells in the portion of the ectoderm (exterior) that will become the mammal's eye undergo differential adhesion, with the result that they form an interior colony and an exterior colony. The interior colony bends to form a hollow niche, which is then "invaded" by the exterior colony. The invading colony, after several further foldings, forms the lens and the cornea; the interior colony forms the retina. (See figure 16.1.) At each stage in this complicated folding process, the niches formed by the dividing cells exchange signals

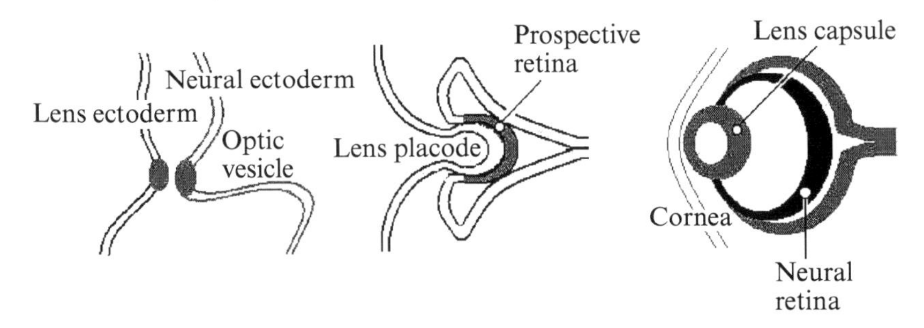

Figure 16.1
A schematic diagram of the morphogenesis of an eye.

that turn genes on and off, thereby modifying adhesion and other parts of the development program.

Unfortunately, the niche concept, though widely used, is typically defined informally or in a narrow, context-dependent way. A well-defined, widely applicable concept of niche should encompass common features of the different versions:

Multiplier effects Resources don't disappear on consumption; they are transformed in passing from agent to agent. As a result, an injection of an additional resource (say, from outside the niche) propagates from agent to agent, multiplying the effects of the injection. (The economic version of this effect was proved by Samuelson.)

Recycling Agents are distributed spatially, and they interact in tangled loops of conditional responses and resource-passing (e.g., phenotypic plasticity in ecosystems (Levin 1999)).

Persistence Often patterns of interaction, imposed on a "flow" of ever-changing agents, persist well beyond the lifetimes of individual agents. These persistent patterns act as niches within which diverse agents can co-exist. The introduction of a new kind of agent to a niche—say, from some other niche (e.g., an invasive species) or by recombination of tags (e.g., a new flu virus)—can cause a torrent of further innovations exploiting the new possibilities.

The concept of a *community* within a network (Newman, Barabasi, and Watts 2006) provides a starting point that leads naturally to an overarching definition of niche. *A niche is a diverse array of agents that regularly exchange resources and depend on that exchange for continued existence.* Most signal/boundary systems exhibit counterparts of ecological niche interactions—symbiosis, mutualism, predation, and the like. From this definition of niche, it is relatively easy to move to an evolutionary dynamic for niches, because the conditional actions that underpin the interlocking activities can be defined and compared in a uniform way by means of tag-sensitive rules. (See chapter 3.)

Defining niches in terms of tag-based rules leads to an important mechanism-oriented question: Can mechanisms for manipulating tags (such as recombination), in combination with selection for the ability to collect resources, lead from simple niches to complex niches? The conditional actions of the agents lead to non-additive effects that cannot be usefully averaged, so a purely statistical approach isn't likely to provide answers to this question. Statistical approaches wind up in the same cul-de-sac as statistical approaches to understanding computer programming. The "trends" suggested by a series of "snapshots" based on an average over agents' activities rarely give reliable predictions or opportunities for control—instead of "clearing" of a market, we get "bubbles" and "crashes."

A more fruitful approach to questions of this kind requires an examination of the coevolution of situated agents—agents that have locations in both a geographic sense and a network sense. The "bandits" discussed in chapter 9 offer a boundary-oriented setting for examining situated agents. Each bandit (slot machine) has a queue of players that share any payoffs that occur. The longer the queue, the less each agent in the queue receives. Each bandit can be looked upon as a tagged urn that dispenses a mixture of colored balls (resources), with agents forming a queue to share those resources. From a niche point of view, an agent's location (situation) is determined by the queue it joins. In the tagged-urn version, an agent can migrate to a queue only if it has a tag that satisfies an entry condition assigned to the queue. With this provision, a "generalist" agent accesses a broad array of urns but often winds up in queues with low payoff per agent because of crowding, whereas a "specialist" agent accesses a queue that is less crowded but more difficult to locate. As was shown in chapter 9, typical niche effects, such as crowding, niche invasion, and specialization, can be examined via the queued bandit approach.

In a more complex, still more niche-like version of the queued bandits, agents within a queue exchange resources with other agents in the queue via the usual tag-based interactions. Displacement of one kind of agent by another can then arise when a specialist reproduces in a given queue at a faster pace than a generalist. The generalist survives only in queues where it isn't out-competed by specialists. Symbiosis arises when two agents make different requirements on the queue's resources and, via tag-based interactions, exchange their surplus shares of particular resources. Groups of agents can also organize into default hierarchies, which leads to more rapid accumulation of adaptations. (See chapter 6.)

16.3 Theory

Progressive co-adaptation of tags causes continual change in tag-based interactions in signal/boundary systems. That is a major complication, but it is also a source of insights. As we have seen, widely different agent-based signal/boundary systems, ranging from biological cells to governments, exhibit the same general features, notably the following:

semi-autonomous subsystems (agents)

hierarchical organization (agents composed of agents)

sustained diversity (agents exploiting different strategies)

extensive recycling of resources (agents converting resources and passing them on).

As was emphasized in the preceding section, it is much easier to compare different signal/boundary systems when we can find a mechanism-oriented framework that encompasses all signal/boundary systems. With such a framework, we can organize the diverse array of observations of signal/boundary systems in a way comparable to the way in which the valence mechanisms suggested by the periodic table organized the endlessly varied observations of the alchemists.

Mechanisms that generate and modify tags via recombination and other evolutionary mechanisms play a pivotal role in the framework proposed here. The definition of fitness is critical when evolutionary mechanisms come into play. The approach taken here emphasizes *implicit* fitness, in which the fitness of an agent depends on its ability to collect the resources necessary to duplicate its structure, as in the Echo models (Holland 1995). For an agent defined via a hierarchy of tagged urns, this requirement entails collecting the "letters" necessary to produce copies of the urns' entry and exit conditions, as well as copies of the "catalyst rules" that determine signal processing within the agent. Both conditions and rules are defined over a three-letter alphabet {1,0,#}, so the agent must collect sufficient numbers of these letters to replicate its structures. It does this by means of its programmed interactions with other agents, and by collecting sources of letters (if any) in its environment. Implicit fitness, so defined, opens the way for a great diversity of agents and niches, with recycling playing a critical role.

Dynamic (finitely) generated systems, discussed in chapter 12 and 14, offer a framework that meets the foregoing desiderata. (I have suppressed 'finitely' in the abbreviation *dgs* to keep it short.) The *dgs* framework is a generalization of the constrained generating procedures developed in Holland 1998. Within the *dgs* framework, different signal/boundary systems are described with the same alphabet and the same fixed set of *dgs* operators (and meta-operators). Each agent is represented by a set of tagged strings defining both its internal hierarchy of boundaries (an urn hierarchy) and the content of the compartments in that hierarchy. Tags, besides identifying parts of the agent, make possible the sequencing of rules, thereby making the agent programmable within the *dgs*. Overall, tags have a role similar to the role of motifs in molecular genetics, determining the specificity of counterparts of enhancers, promoters, silencers, ligands, catalysts, and the like (Alberts 2007).

In a *dgs,* the generated succession of sets (populations) of tagged strings represents successive generations of agents. Operators and meta-operators—particularly meta-operators acting as a genetic algorithm—modify the tags, thereby modifying the urns and their content. The result is a trajectory of progressive adaptations and coevolution in successive generations. Everything within the *dgs*, whatever the signal/boundary system being modeled, uses the same operators and meta-operators to manipulate the strings. As a consequence, the coevolution of signals and boundaries in quite different signal/boundary systems can be compared within the *dgs*. The only differences between *dgs* models of different signal/boundary systems arise from differences in the starting set of strings (the X(0) of chapter 14). The corpus generated from X(0)—the sequence of generations—follows entirely from the actions of the fixed set of operators and meta-operators, much as theorems are generated from different axiom sets use the same fixed rules of deduction.

The procedures for developing a *dgs* model relevant to a particular set of signal/boundary observations are similar to the procedures for writing a computer simulation. That is, the *dgs* generators and operators serve much like the instruction set for a computer. However, the *dgs* generators and operators are selected to provide direct descriptions of the rules and conditions employed by tagged urns. As a result, networks of urn and agent interactions become important emergent phenomena in a *dgs*. Moreover, network modifications are *generated* by *dgs* operators that produce agent adaptations. Coevolutionary changes in tags can generate a cascade of subtle network changes that are difficult to characterize in terms of direct manipulation of nodes and edges. Roughly, the difference between emergent modification and direct modification is the difference between modifying the genes that specify a metabolic network and directly modifying the reactants in that network.

Tags—in the guise of motifs, active sites, conditions, and the like—have a large role in determining signal/boundary dynamics. For that reason, a relevant *dgs* theory should provide ways of generating new tags from extant tags. In a *dgs*, a building block for a tag is a short string that appears at the same loci in several different tags (akin to a motif in genomics). In the terminology of genetic algorithms, such building blocks are called *schemata*. Combining a small number of such building blocks produces a wide variety of "plausible" new tags for defining conditions and signals. Some of these "plausible combinations" facilitate the formation and testing of new default hierarchies, "production lines," and the like. Because populations are involved, many variants can be tried simultaneously without significant loss of capability.

16.4 Mathematical models

Theory is used to suggest experiments and "laws" that explain features appearing repeatedly in some wide range of observations. The *dgs* framework proposed in this book aims at the pervasive occurrence of hierarchical signal/boundary interactions in complex adaptive systems. To provide the abstractions necessary for theory, the framework centers on tagged urns that serve as semi-permeable boundaries for agent compartments. The compartments, in turn, contain reactions (and interactions) defined by a "billiard-ball chemistry." Under this regime, the state of this *dgs* model at any time is a probability distribution over the possible concentrations of signals and resources within the tagged urns. The probabilities defined by this framework open the way for using the mathematics of Markov processes. (See chapter 15.)

Markov matrices nicely capture the conditional probabilities that define the changes imposed by diffusion/reaction systems, as represented by diffusion and catalyzed reactions in the tagged-urn models. Each entry in the matrix gives the conditional probability P(c′|c) of going from one arrangement of balls in the urns (c, the row label) to another arrangement (c′, the column label). Because the system has a finite number of urns, the number of possible arrangements (v) is finite, so there are v rows and columns in the matrix, and there v components in the state vector that gives the probability of any given arrangement.

Networks offer another theoretical approach, one that is more "pictorial." Each node c in a network represents a particular arrangement of balls among urns, and two nodes c and c' are connected by a directed edge just in case moving a single ball can take the system from node (arrangement) c to node c'. When the edges are weighted with the corresponding probabilities, the network provides the same information as the Markov matrix.

With the help of these theoretical devices, we can approach a question of some importance in evolution of signal/boundary systems. Under what conditions does the system "settle down" (that is, reach an equilibrium)? When balls move randomly (diffuse) among a fixed set of urns with no entry and exit conditions, balls of any given color are equally distributed among the urns. How do entry and exit conditions on the urns modify this result? Standard methods for Markov theory can be used to show that, under conditions usually satisfied by balls diffusing among the tagged urns, the probability of each arrangement becomes fixed after sufficient time elapses. More technically, the probability vector X(t) describing the state of the signal/boundary system settles down to a fixed point. That is, X(t + 1) = X(t)M = X(t), where M is the matrix describing the diffusion process.

However, most complex adaptive systems don't settle down. The adaptations and coevolution produced by mechanisms such as genetic operators (recombination, mutation, etc.) change tags and tagged urns more or less rapidly, thereby modifying the Markov matrix M. Nevertheless, a look at the steady states before and after one or two applications of genetic operators can give some idea of the resulting changes in fitness. And, as was pointed out earlier, the changes that produce higher fitness are the ones that persist.

In considering mathematical approaches to signal/boundary systems, it is important to recognize that agents situated in a physical environment *do* encounter regularities. That is, the environments encountered are a highly constrained subset of the set of all conceivable environments. It is a bad mistake to assume that nearly all conceivable environments occur with non-zero probability. That mistake leads to misconceptions such as Laplace's calculation that the sun will rise tomorrow or the "no free lunch" theorems that assert that adaptive algorithms offer no overall advantage (Wolpert 1992). Indeed, it is these observed regularities that make the tiered laws of the physical sciences possible.

In realistic environments, regularities offer exploitable opportunities for the formation of niches involving persistent, interdependent interactions of diverse agents. Once a niche is formed through adaptation and coevolution, its persistence depends on using arrangements of signals and boundaries to exploit environmental regularities. In the signal/boundary context, recombination of tags offers substantial advantages over uniform random search or searches that depend on the law of large numbers. Agents that exploit the tags with conditional rules play a pivotal role in sustaining the niche. And while the niche persists, new opportunities for interaction proliferate, offering roles for still newer agents. (See section 16.1.)

16.5 Some questions

Because bounded subsystems (especially adaptive agents) are ubiquitous features of signal/boundary systems, priority goes to questions concerning their origin, their interaction, and their coevolution. As was emphasized above, such questions center on the coevolution of tags that mediate the routing and collecting of signals within and between agents. Coevolution, then, centers on various kinds of recombination, ranging from somatic recombination in the immune system, through various kinds of horizontal transfer in viruses and prokaryotes, to crossover in eukaryotes, and technological innovation via new combinations of tested components. Answers to two broad questions would greatly improve our understanding of this coevolutionary process.

The first of these questions is "What are typical steps in the formation and evolution of niches?" There are several closely related questions: How is the development of niches affected by agents' capabilities, such as conditional response (e.g., phenotypic plasticity), anticipation, and "lookahead" based on internal models? What are the mechanisms that implement the transition from individual "generalists" to networks of "specialists"? What kinds of niches promote robustness in signal/boundary networks? What kinds of niches encourage increasing diversity?

The second question is "What are the mechanisms that favor the boundary hierarchies?" Again there is a closely related question: Can changes in the networks produced by coevolutionary processes be described in terms of general operations on nodes and edges?

It is a thesis of this book that general answers to these questions hinge on constructing a *dgs* based on operators and generators extracted from the commonalities of signal/boundary system hierarchies.

16.6 A short version of the short version

Complex adaptive systems—cells, rainforests, markets, language, and the Internet, to name a few—are characterized by complex, ever-changing interactions of signals and boundaries. This book provides exploratory models for examining the adaptations generated by signal/boundary coevolution. The exploration centers on two kinds of models: adaptive agents and tagged urns. Adaptive agents (defined by signal-processing rules that provide for parallel processing and adaptation under a genetic algorithm) model the evolution of hierarchical systems that employ many signals (resources) interacting simultaneously. Tagged urns (modifications of the urns used in probability theory) use entry and exit conditions to control the flow of balls (signals) between urns, thus providing explicit formal models of semi-permeable boundaries. In these models, both signals and boundaries are constructed from building blocks provided by tags and parts of tags—the counterparts of active sites, motifs, and message headers. The framework of dynamic generated systems combines adaptive-agent models and tagged-urn models to provide a precise formalism for exploring the evolution of signal/boundary systems.

References

Alberts, B., et al. 2007. *Molecular Biology of the Cell*. Garland.

Allen, M. 1977. *Darwin and His Flowers: The Key to Natural Selection*. Taplinger.

Arthur, W. B. 2009. *The Nature of Technology*. Free Press.

Arthur, W. B., et al. 1997. Asset pricing under endogenous expectations in an artificial stock market. *Economic Notes* 26 (2): 297–330.

Ashby, W. R. 1952. *Design for a Brain*. Wiley.

Asimov, I. 1988. *Understanding Physics*. Buccaneer.

Axelrod, R., and Cohen, M. D. 1999. *Harnessing Complexity*. Free Press.

Babbage, C., and Babbage, H. P. 2010. *Babbage's Calculating Engines: Being a Collection of Papers Relating to Them*. Cambridge University Press.

Birkoff, G., and MacLane, S. 2008. *A Survey of Modern Algebra*. A K Peters.

Braitenberg, V. 1984. *Vehicles: Experiments in Synthetic Psychology*. MIT Press.

Buss, L. W. 1987. *The Evolution of Individuality*. Princeton University Press.

Bybee, J. 2006. *Frequency of Use and the Organization of Language*. Oxford University Press.

Chomsky, N. 1965. *Aspects of a Theory of Syntax*. MIT Press.

Christiansen, F. B., and Feldman, M. W. 1986. *Population Genetics*. Blackwell.

Cole, L. C. 1954. Some features of random population cycles. *Journal of Wildlife Management* 18 (1): 2–24.

Davidson, E. H. 2006. *The Regulatory Genome*. Academic Press.

Dennett, D. C. 1992. *Consciousness Explained*. Little, Brown.

Dobzhansky, T., et al. 1977. *Evolution*. Freeman.

Epstein, J. M., and Axtell, R. L. 1996. *Growing Artificial Societies*. MIT Press.

Felicity, A., Clements, A., Webb, C., and Lithgow, T. 2010. Tinkering inside the organelle. *Science* 327: 649–650.

Feller, W. 1968. *An Introduction to Probability Theory and Its Applications*, volume 1. Wiley.

Feynman, R. P., Leighton, R. B., and Sands, M. 2005. *The Feynman Lectures on Physics Including Feynman's Tips on Physics*. Addison-Wesley.

Fisher, R. A. 1930. *The Genetical Theory of Natural Selection*. Clarendon.

Five Graces Group (C. Beckner et al.). 2009. Language is a complex adaptive system. *Language Learning* 59: 1–26.

Fontana, W. 2006. Pulling strings. *Science* 314: 1552–1553.

Forgacs, G., and Newman, S. A. 2005. *Biological Physics of the Developing Embryo*. Cambridge University Press.

Forsyth, A., and Miyata, K. 1984. *Tropical Nature*. Scribner.

Gao, H. 2001. *The Physical Foundation of the Patterning of Physical Action Verbs*. Lund University Press.

Gao, H., and Holland, J. H. 2008. *Agent-Based Models of Levels of Conscious*. SFI Working Papers.

Gilbert, S. F., and Epel, D. 2009. *Ecological Developmental Biology*. Sinauer.

Goldberg, D. E. 2002. *The Design of Innovation*. Kluwer.

Good, M. C., Zalatan, J. G., and Lim, W. A. 2011. Scaffold proteins: Hubs for controlling the flow of cellular information. *Science* 232: 680–686.

Han, J., Li, M., and Guo, L. 2006. Soft control on collective behavior of a group of autonomous agents by a shill agent. *Journal of Systems Science and Complexity* 19 (1): 54–62.

Hebb, D. O. 1949. *The Organization of Behavior*. Wiley.

Hofstadter, D. R. 1999. *Gödel, Escher, Bach: An Eternal Golden Braid*. Basic Books.

Holland, J. H., Holyoak, K. J., Nisbett, R. E., and Thagard, P. R. 1986. *Induction: Processes of Inference, Learning, and Discovery*. MIT Press.

Holland, J. H. 1992. *Adaptation in Natural and Artificial Systems*. MIT Press.

Holland, J. H. 1995. *Hidden Order: How Adaptation Builds Complexity*. Addison-Wesley.

Holland, J. H. 1998. *Emergence: From Chaos to Order*. Addison-Wesley.

Holland, J. H. 2002. Exploring the evolution of complexity in signaling networks. *Complexity* 7 (2): 34–45.

Holland, J. H., Tao, G., and Wang, W. S.-Y. 2005. Co-evolution of lexicon and syntax from a simulation perspective. *Complexity* 10: 50–62.

Holldobler, B., and Wilson, E. O. 1990. *The Ants*. Harvard University Press.

Ke, J., and Holland, J. H. 2006. Language origin from an emergentist perspective. *Applied Linguistics* 27: 691–716.

Kemeny, J. G., and Snell, J. L. 1976. *Finite Markov Chains*. Springer.

Kleene, S. C. 1956. Representation of events in nerve nets and finite automata. In *Automata Studies*, ed. C. Shannon and J. McCarthy. Princeton.

Konig, D. 1936. *Theorie der Graphen*. Akademische Verlag.

Krugman, P., Wells, R. and Graddy K. 2010. *Essentials of Economics*. Worth.

Lanzi, P. R., et al. 2000. *Learning Classifier Systems*. Springer.

Levin, S. A. 1999. *Fragile Dominion*. Perseus.

Lindsley, D. L., and Grell, E. H. 1967. *Genetic Variations of* Drosophila Melanogaster. Carnegie Institution of Washington.

Maxwell, J. C. (ed. W. Davidson). 1890. *The Scientific Papers of James Clerk Maxwell*. Cambridge University Press (BiblioLife).

McClelland, J. L., and Rumelhart, D. E. 1986. *Parallel Distributed Processing: Explorations in the Microstructure of Cognition*. MIT Press.

Mitchell, M. 1996. *An Introduction to Genetic Algorithms*. MIT Press.

Mitchell, M. 2009. *Complexity A Guided Tour*. Oxford University Press.

Newman, J. R. 2003. *The World of Mathematics*. Simon & Schuster.

Newman, M., Barabasi, A.-L., and Watts, D. J. 2006. *The Structure and Dynamics of Networks*. Princeton University Press.

Nowak, M. A. 2006. *Evolutionary Dynamics*. Harvard University Press.

Rapoport, A. 1960. *Fights, Games, and Debates*. University of Michigan Press.

Rochester, N., et al. 1988. 1955 Tests on a cell assembly theory of the action of the brain. In *Neurocomputing*, ed. J. Anderson and E. Rosenfeld. MIT Press.

Samuelson, P., and Nordhaus, W. 2009. *Economics*. McGraw-Hill.

Simon, H. A. 1996. *The Sciences of the Artificial*. MIT Press.

Slack, E., et al. 2009. Innate and adaptive immunity cooperate flexibly to maintain host-microbiota mutualism. *Science* 325: 617–620.

Smith, A. 1776. *The Wealth of Nations*. Penguin.

Smith, E., and Morowitz, H. J. 2004. Universality in intermediary metabolism. *Proceedings of the National Academy of Sciences of the United States of America* 101: 13168–13173.

Steels, L., and Kaplan, F. 2002. 2002 Bootstrapping grounded word semantics. In *Linguistic Evolution through Language Acquisition*, ed. E. Briscoe. Cambridge University Press.

Tarski, A., Mostowski, A., and Robinson, R. M. 2010 *Undecidable Theories: Studies in Logic and the Foundation of Mathematics.* Dover.

Tolstoy, I. 1981. *James Clerk Maxwell.* University of Chicago Press.

Turing, A. M. 1936. On computable numbers, with an application to the Entscheidungsproblem. *Proceedings of the London Mathematical Society* 42: 230–265.

von Neumann, J. 1966. In *Theory of Self-Reproducing Automata*, ed. A. Burks. University of Illinois Press.

Walker, P. M. B., ed. 1990. *Cambridge Dictionary of Biology*. Cambridge University Press.

Whitehead, A. N., and Russell, B. 1910. *Principia Mathematica.* Cambridge University Press.

Wolpert, D. H. 1992. On the connection between in-sample testing and generalization error. *Complex Systems* 6: 47–94.

Zelazo, P. D., Gao, H. H., and Todd, R. 2007. Development of consciousness. In *The Cambridge Handbook of Consciousness*, ed. P. Zelazo et al. Cambridge University Press.

Index